International Trade and Climate Change

ENVIRONMENT AND DEVELOPMENT

A fundamental element of sustainable development is environmental sustainability. Hence, this series was created in 2007 to cover current and emerging issues in order to promote debate and broaden the understanding of environmental challenges as integral to achieving equitable and sustained economic growth. The series will draw on analysis and practical experience from across the World Bank and from client countries. The manuscripts chosen for publication will be central to the implementation of the World Bank's Environment strategy, and relevant to the development community, policy makers, and academia. In that spirit, forthcoming volumes are expected to address environmental health, natural resources management, strategic environmental assessment, policy instruments, and environmental institutions.

Second volume in the series
Proverty and the Environment: Understanding Linkages at the Household Level

International Trade and Climate Change

Economic, Legal, and Institutional Perspectives

THE WORLD BANK
Washington, DC

1818 H Street, NW
Washington, DC 20433
Telephone 202-473-1000
Internet www.worldbank.org
E-mail feedback@worldbank.org

1 2 3 4 :: 10 09 08 07

ISBN: 978-0-8213-7225-8
e-ISBN: 978-0-8213-7226-5
DOI: 10.1596/ 978-0-8213-7225-8

Design: Auras Design
Cover photos:
Windmills: © Paul Giamou/Getty Images
Child in floodwater: © Gideon Mendel/Corbis

Library of Congress Cataloging-in-Publication Data has been applied for.

CONTENTS

Figures

Tables

Acknowledgments

This study is the product of a team composed of Muthukumara Mani (Task Team Leader and Senior Environmental Economist, ENV/World Bank), Chandrasekar Govindarajalu (Senior Environmental Specialist, ENV/World Bank), Hiau Looi Kee (Senior Economist, DECRG/World Bank), Sunanda Kishore (Consultant, ENV/World Bank), Eri Tatsui (Consultant, ENV/World Bank), Cizuka Seki (Consultant, ENV/World Bank), Sachiko Morita (Consultant, LEGEN/World Bank), and Mahesh Sugathan (Program Coordinator, Economics and Trade Policy/ICTSD).

In preparing this study, the team has greatly benefited from detailed comments by peer reviewers: Bernard Hoekman (Senior Advisor, DECRG/World Bank), Thomas Brewer (Associate Professor, Georgetown University), Richard Damania (Senior Environmental Economist, SASEN/World Bank), Donald Larson (Senior Economist, DECRG/World Bank), and Masaki Takahashi (Senior Power Engineer, ETWEN/World Bank).

In addition, the following people provided valuable inputs and written comments to the team: Kirk Hamilton (Team Leader and Lead Environmental Economist, ENV/World Bank), Charles E. Di Leva (Chief Counsel, LEGEN/World Bank), Laura Tlaiye (Sector Manager, ENV/World Bank), Giovanni Ruta (Economist, ENV/World Bank), Sushenjit Bandyopadhyay (Senior Environmental Economist, ENV/World Bank), Anil Markandya (Professor, University of Bath) and Moustapha Kamal Gueye (Senior Programme Manager, Environment Cluster/ICTSD). The team greatly appreciates the contribution and guidance on technical issues especially pertaining to the WTO from ICTSD. Editorial support was provided by Alexandra Sears and Robert Livernash. The Environment Sector Manager is Laura Tlaiye, and the Environment Department Director is James Warren Evans.

Abbreviations

AHTN	ASEAN Harmonized Tariff Nomenclature
APEC	Asia Pacific Economic Community
AVE	ad valorem equivalent
BAU	business as usual
CCS	carbon capture and storage
CDM	Clean Development Mechanism
CFL	compact fluorescent lamp
CHP	combined heat and power
COP	Conference of Parties
CTE	Committee on Trade and Environment
DNA	designated national authority
DOE	Department of Energy
DSM	demand-side management
EC	European Commission
EEA	European Environment Agency
EEC	European Economic Community
EGS	environmental goods and services
EIT	economies in transition
ELI	Efficient Lighting Initiative
EPP	environmentally preferable product
EST	environmentally sound technologies
ETS	Emissions Trading Scheme
EU	European Union
EU ETS	European Union Emission Trading System
FDI	foreign direct investment
FGG	flue gas desulfurization
FTA	free trade agreement
GATT	General Agreement on Tariffs and Trade
GDP	gross domestic product
GE	General Electric
GHG	greenhouse gas
GPA	government procurement agreement
GT	gigaton
GW	gigawatt
GWEC	Global Wind Energy Council
HS	harmonized system
IEA	International Energy Agency
IGCC	integrated (coal) gasification combined cycle

IJV	international joint venture
IPCC	Intergovernmental Panel on Climate Change
IPR	intellectual property rights
ISIC	International Standard of Industrial Classification of All Economic Activities
ITA	information technology agreement
JI	joint implementation
KWH	kilowatt-hour
LED	light emitting diode
LPG	liquefied petroleum gas
MEA	multilateral environmental agreement
MEPS	minimum energy performance standards
MFN	most-favored nation
MNE	multinational enterprise
MW	megawatt
NAFTA	North American Free Trade Agreement
NOx	nitrogen oxide
NTB	nontariff barrier
OECD	Organisation for Economic Co-operation and Development
PC	PetCoke
PPM	parts per million
PPM	process and production methods
PURPA	Public Utility Regulatory Policies Act of 1978
PV	photovoltaic
R&D	research and development
RPS	renewable portfolio standards
SDN	sustainable development network
TRC	tradable renewable energy certificate
TRIMS	trade-related investment measures
TWH	terawatt-hour
UNCTAD	United Nations Conference on Trade and Development
UNEP	United Nations Environment Programme
UNFCCC	United Nations Framework Convention on Climate Change
USTR	United States Trade Representative
VA	voluntary agreement
VAT	value-added tax
WCO	World Customs Organization
WITS	World Integrated Trade Solution
WMO	World Meteorological Organization
WTO	World Trade Organization

Measurements: All currency is in U.S. dollars. All tons are metric tons.

CHAPTER 1

Introduction and Overview

THE ECONOMIC, SOCIAL, AND DEVELOPMENTAL consequences of climate change have received increasing recognition worldwide. The *Stern Review* (2006) notes that climate change is a serious and urgent problem, global in its cause and consequences. Current actions are not enough if we are to stabilize greenhouse gases (GHGs) at any acceptable level. The economic challenges are complex and will require a long-term international collaboration to tackle them. The recent report of the Intergovernmental Panel on Climate Change (IPCC) also categorically states that the impacts of climate change will vary regionally, but aggregated and discounted to the present, they are very likely to impose net annual costs that will increase over time as global temperatures increase (IPCC 2007). The Kyoto Protocol remains the key international mechanism under which the industrial countries have committed to reduce their emissions of carbon dioxide and other greenhouse gases (see box 1.1).

A number of issues still need to be resolved with regard to the efficient implementation of emissions reduction goals. Although 172 countries and a regional economic integration organization (the European Economic Community) are parties to the agreement (representing over 61 percent of emissions), only a few industrialized countries are actually required to cut their emissions (see appendix 1 in this report for a list of Kyoto Protocol signatories and their emission targets). The United States, which is the world's largest emitter, and Australia have not ratified the

BOX 1.1

The Kyoto Protocol

The Kyoto Protocol to the United Nations Framework Convention on Climate Change (UNFCCC) entered into force on February 16, 2005, following ratification by Russia. As of May 11, 2007, 172 countries and the regional economic integration organization (European Economic Community) have ratified, accepted, approved, or acceded to the Kyoto Protocol. The UNFCCC includes the principle of "common but differentiated responsibilities." Under the principle, as stipulated in Article 3, paragraph 1, of the UNFCCC, the parties agreed that (i) the largest share of historical and current global emissions of greenhouse gases has originated in developed countries; (ii) per capita emissions in developing countries are still relatively low; and (iii) the share of global emissions originating in developing countries will grow to meet their social and development needs.

Under the Kyoto Protocol, industrialized countries (called Annex I countries) have to reduce their combined emissions to 5 percent below 1990 levels in the first commitment period of 2008–12. Annex I countries include the industrialized countries that were members of the Organisation for Economic Co-operation and Development (OECD) in 1992, plus countries with economies in transition (the EIT parties), including the Russian Federation, the Baltic states, and several Central and Eastern European states. Countries that have accepted greenhouse gas emissions reduction obligations must submit an annual greenhouse gas inventory. Non-Annex I countries (developing countries) that have ratified the Protocol do not have to commit to specific targets because they face potential technical and economic constraints. Nevertheless, they have to report their emissions levels and develop national climate change mitigation programs.

Although the average emissions reduction is 5 percent, each country agreed to its own specific target. Within the Annex I countries, differentiated national targets range from 8 percent reductions for the European Union (EU) to a 10 percent allowable increase in emissions for Iceland.

Further, while Annex I countries must put in place domestic policies and measures to achieve their targets, the Protocol does not oblige governments to implement any particular policy, instead allowing countries to seek optimal ways to achieve greenhouse gas emissions reduction and to adjust their climate change strategies to the circumstances of their economies. The Protocol defines three flexibility mechanisms to help Annex I parties lower the overall costs of achieving emissions targets. The three mechanisms—Joint Implementation (JI), the Clean Development Mechanism (CDM), and emissions trading—allow them to reduce emissions, or increase greenhouse gas removals, in other countries, where it can be done more cheaply than at home.

Source: UNFCCC, Essential Background, http://unfccc.int/essential_background/items/2877.php.

Protocol. The United States has conditioned its entry on further engagement of major developing country emitters, such as China and India.

In countries that have begun to implement the Kyoto regime, this disparity in commitments has fueled a debate on issues of competitiveness and other economic impacts.[1] Businesses in many Kyoto-implementing countries have already started to urge their governments to ease competitive pressures through measures such as a border tax. A recent European Commission report suggests taxing goods imported from countries that do not impose a CO_2 cap on their industry as a way to compensate for the costs of climate change measures. Stiglitz (2006) advocates that Europe, Japan, and others adhering to the Kyoto Protocol should restrict or tax the import of American goods to make up for the fact that U.S. producers do not incur GHG-related costs of production and, therefore, produce goods that are less responsible toward the environment.

Unlike some other global environmental treaties—such as the Montreal Protocol on Substances that Deplete the Ozone Layer—the Kyoto Protocol does not contain explicit trade measures to enforce compliance.[2] Nor does it stipulate specific methods by which the members should design and implement policies to address climate change commitments. Nevertheless, as this study demonstrates, the disparity in effort between developed countries is leading to concerns about competitiveness and principles of equity. In turn, these concerns lead to much speculation about whether Kyoto should develop trade sanctions, or whether other Kyoto-supportive trade measures are appropriate to protect those industries that are absorbing the cost of GHG-reducing technologies. As a result, there is additional speculation about a potential conflict between the Kyoto and WTO regimes (Brewer 2003; Georgieva and Mani 2006; Loose 2001).

Reducing emissions in industrial countries is just one side of the story. It is becoming increasingly clear that developing countries will drive the future of global economic growth. Estimates show that by 2030, about half or more of the purchasing power of the global economy will stem from the developing world. Their share in world GDP could reach 60 percent in terms of purchasing power parity and their share in world trade almost 50 percent (World Bank 2007b). These increases have important implications for both GHG emissions and any future climate regime.

Though developed countries remain the largest per capita emitters of greenhouse gases today, the growth of carbon emissions in the next decades will come primarily from developing countries, which are following the same energy- and carbon-intensive development path as their rich counterparts have done. Among the developing countries, the greatest increase in carbon emissions will emanate from China and India because of their size and growth. It is projected that, between 2020 and 2030, developing country emissions of carbon from energy use will exceed those of developed countries. Any kind of post-Kyoto international regime that will emerge to address climate change cannot ignore these startling facts.

FIGURE 1.1
CO_2 Emissions from Energy Use, 2002-30

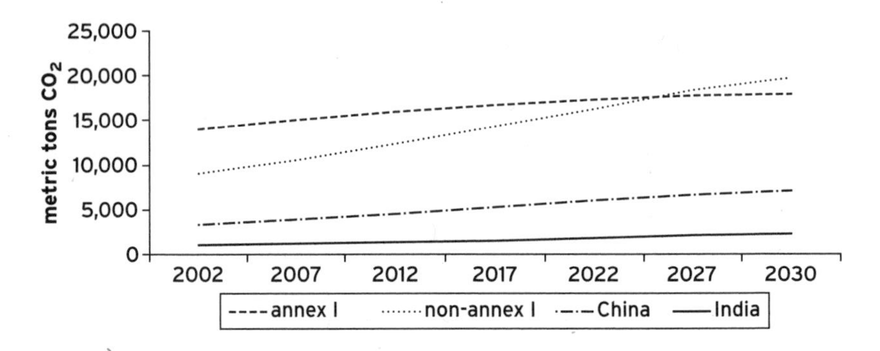

Source: IEA Database 2006.

Climate change is a global challenge requiring international collaborative action. Another area in which countries have successfully committed to a long-term multilateral resolution is the liberalization of international trade. Integration into the world economy has proved to be a powerful means for countries to promote economic growth, development, and poverty reduction. Some developing countries have opened their own economies to take full advantage of the opportunities for economic development through trade, but many have not. The ongoing Doha "Development" Round is seen by many as a potential vehicle for real gains for all economies, and particularly developing economies, in the areas of agricultural reform, improved market access for goods and services, and clarification and improvement of trade disciplines.

The broad objectives of the betterment of current and future human welfare are thus shared by both global trade and climate regimes. Yet both climate and trade agendas have evolved largely independently through the years, despite their mutually supporting objectives and the potential for synergies discussed in this study. While the implementation of the Kyoto Protocol may have brought to light some inherent conflicts between economic growth and environmental protection, the objectives of Kyoto also provide an opportunity for aligning development and energy policies in such a way that they could stimulate production, trade, and investment in cleaner technology options. Since global emission goals and global trade are policy objectives shared by most countries and nearly all of the World Bank's clients, it makes sense to consider the two sets of objectives together.

Technology Options to Stabilize Greenhouse Gas Emissions

The stabilization of GHG concentrations—to as low as 450 ppm CO_2-equivalent—can be achieved by deploying currently available technologies and technologies that are expected to be commercialized in the coming decades in the energy supply, transport, buildings, industry, agriculture, forests, and waste management sectors (IPCC 2007).

In the global discourse on climate change, technologies that help in mitigating the impacts by reducing the GHG emissions have been termed variously as "environmentally sustainable technologies," "environmentally sound technologies," "sustainable energy technologies," "clean energy technologies" (used in this report), and several other terms. Substantively there is little difference in the core set of technologies—energy efficiency, renewable energy, and a few other high-GHG-impact technologies—these technologies represent an evolution of a global discourse on the topic of climate change and the political realities of the stakeholders. The availability of these climate-friendly technologies is critical if developing countries' are to achieve low-carbon growth paths.

In the recent literature, Socolow and others (2004) have used these technologies to identify strategies that are climate friendly. They introduce the concept of "stabilization wedges," which is helpful in understanding the scale of the challenge in order to stabilize carbon emissions by 2054—aiming at a CO_2 atmospheric concentration of 500 ppm. Each wedge results in a reduction in the rate of carbon emissions of 1 billion tons of carbon per year by 2054, resulting in 25 billion tons over 50 years. In other words, each wedge has the potential to reduce emissions by an increasing amount per year, starting at very low levels now and reaching 1 gigaton (Gt) per year by 2054, by which time emissions of CO_2 will have been reduced by a cumulative 25 Gt.

The Socolow study examined 15 such strategies, each based on a known technology with a potential to contribute to carbon mitigation (box 1.2). For example, a wedge from renewable electricity replacing coal-based power is available from a 50-fold expansion of wind by 2054 or a 700-fold expansion of solar photovoltaics relative to today.

More recently, the IPCC Working Group III (IPCC 2007) also called for a mix of policy instruments and incentives to reduce GHG emissions to a manageable 450 ppm. Specifically, the report suggests the following:

- Policies that provide a real or implicit price of carbon could create incentives for producers and consumers to significantly invest in low-GHG products, technologies, and processes, including economic instruments, regulation (e.g., standards), and government funding and tax credits. Integrating climate policies into broader development policies would facilitate the transition to a low-carbon economy.
- It is economically feasible to halt, and possibly reverse, the growth in global GHG emissions in order to stabilize their atmospheric concentrations. Key mitigation technologies and practices projected to be commercialized before 2030 include carbon capture and storage, advanced nuclear power, renewable

BOX 1.2

Summary List of Technologies Considered as "Wedges" for Climate Change Mitigation

1. **End-user efficiency and conservation**
   - Increase fuel economy of automobiles
   - Reduce automobile use by telecommuting, mass transit, urban design
   - Reduce electricity use in homes, offices, and stores
2. **Power generation**
   - Increase efficiency of coal-fired plants
   - Increase gas baseload power (reduce coal baseload power)
3. **Carbon capture and storage (CCS)**
   - Install CCS at large, baseload coal-fired plants
   - Install CCS at coal-fired plants to produce hydrogen for vehicles
   - Install CCS at a coal-to-synfuels power plant
4. **Alternative energy sources**
   - Increase nuclear power (reduce coal)
   - Increase wind power (reduce coal)
   - Increase photovoltaic power (reduce coal)
   - Use wind to produce hydrogen for fuel cell cars
   - Substitute biofuels for fossil fuels
5. **Agriculture and fisheries**
   - Reduce deforestation, increase reforestation and afforestation, add plantations
   - Increase conservation tillage in cropland

Source: Socolow and others 2004.

energy (e.g., tidal and wave energy), second-generation biofuels, advanced electric and hybrid vehicles, and integrated design of commercial buildings.

- Governments must invest more in energy research and development (R&D) to deliver low-GHG technologies.

Successful GHG mitigation approaches, however, need to support developing countries' economic and social development needs and institutional, financial, and technical capacity. These countries cannot take on the same commitments as the developed countries as they often lack institutional, financial, and technical capacity, which will influence their ability to implement and comply with climate commitments.

In addition, developing countries must deal with poverty and other social challenges, and they may be reluctant to adopt restrictive policies that could limit economic growth and pose any threat to energy security. As a result, climate change may rank as a low political priority. However, developing countries are also more vulnerable to the impacts of climate change. Their economies are more dependent on climate-sensitive sectors such as agriculture and forestry, and they lack the infrastructure or resources to respond to the results of changes in climate. Hence, any market-driven mechanism that facilitates the transfer of clean technology—at the same time entailing minimal costs to the developing countries' economies—may be viewed more favorably than the more traditional command-and-control regimes.

Technology transfer to developing countries has been a mandate of the UNFCCC. The convention includes provisions calling for the transfer to developing countries of technology and know-how related to environmentally sound technologies, or ESTs (Article 4, paragraphs 5 and 7).[3] The convention's component on enabling environments specifically focuses on government actions—such as fair trade policies; removal of technical, legal, and administrative barriers to technology transfer; sound economic policies; regulatory frameworks; and transparency—that create an environment conducive to private and public sector technology transfer.

Various sessions of the Conference of Parties (COP) have discussed this issue and have made decisions to promote development and transfer of ESTs. A key milestone in this regard was achieved at the COP-7 in Marrakesh in 2001, when a technology transfer framework was adopted to enhance implementation of climate-friendly technologies.

The *Stern Review* (2006) on the economics of climate change also identifies the transfer of energy-efficient and low-carbon technologies to developing countries as key to reducing the energy intensity of production. It further observes that "the reduction of tariff and nontariff barriers for low-carbon goods and services, including within the Doha Development Round of international trade negotiations, could provide further opportunities to accelerate the diffusion of key technologies" (p. xxv).

In that context, this study addresses an important policy question: how changes in trade policies and international cooperation on trade policies can help address global environmental spillovers, especially GHG emissions, and what the potential effects of national environmental policies aimed at global environmental problems might be for trade and investment.

The Debate on Trade and the Environment Revisited

There has been much debate over the last two decades on the role international trade plays in determining environmental outcomes. This has led to both theoretical work, identifying a series of hypotheses linking openness to trade and environmental quality, and empirical work, trying to disentangle some of the suggested linkages using cross-country or within-country data. Much of the focus, however, has been on local pollution issues. Studies have primarily looked at how changes in production and trade flows have altered the pollution intensity of production (composition effect) in both developed and developing countries, and how trade flows may themselves be affected by the level of abatement costs or strictness of pollution regulation in the trading partner countries.[4] A number of more recent studies have looked at the interface of trade and political economy issues and their implications for the environment and natural resources (see box 1.3 for a synopsis of the general debate on trade and the environment).

In the policy arena, the importance of establishing coherent relationships between the trade obligations set out in various bilateral and multilateral trade agreements and environmental policies of countries is now well recognized. Environmental provisions in the General Agreement on Tariffs and Trade (GATT) allow adoption of product-related measures in certain situations if they are "necessary to protect human, animal or plant life or health," or "relat[e] to the conservation of exhaustible natural resources." In addition, other trade agreements—such as NAFTA and the U.S.-Singapore free trade agreement—include provisions that directly address environmental concerns.

Interestingly, the trade-environment debate has so far considered little in terms of global-scale environmental problems—climate change, declining biodiversity, the depletion of ocean fisheries, and the overexploitation of shared resources. These "public goods" issues, which require international cooperative action, can potentially lead to trade tensions if some countries get a "free ride" on the environmental efforts of others.[5] Although mechanisms such as the Kyoto Protocol (and other multilateral environmental agreements) deal with global environmental issues, none of the agreements have universal membership. This imbalance could lead to potential conflicts as treaty-member countries adopt measures to comply with the global agreements, which could be made binding on WTO members who are not parties to the same treaties.

Although there is potential for conflict between trade and the emerging global environmental regime to combat climate change, some issues currently on the agenda of the WTO could potentially be harnessed to promote broader global environmental objectives. For example, a multilateral liberalization of renewable energy sources or an agreement to remove fossil fuel subsidies would equally serve climate change objectives. The WTO negotiations on environmental goods and services could be used as a vehicle for broadening trade in cleaner technology options and thereby help developing countries reduce their greenhouse gas emissions and adapt to climate change. A more transparent and justifiable labeling and standards regime could similarly serve the interests of both trade and global environmental objectives. In addition, more uniform pricing of energy under the UNFCCC could negate some trade issues regarding competitiveness and leakage.

Focus and Results of This Study

In the context of the implications of linkages between trade and climate change, this study assesses the following:

- What are the main policy prescriptions employed by OECD countries to reduce greenhouse gases, and how do they affect the competitiveness of their energy-intensive industries?

BOX 1.3

Environmental Aspects of Bilateral and Multilateral Trade Agreements

The concerns with environmental implications of trade involve both the domestic implications of policy reforms and the global environmental dimension of bilateral and multilateral trade agreements. Although liberalizing reforms generally promote more-efficient resource use (including use of environmental resources), in practice there is no clear-cut reason to expect that trade liberalization will be either good or bad for the environment. The reason is that trade reforms undertaken in the presence of existing market, policy, or institutional imperfections in the environment or natural resource sector may lead to adverse environmental impacts. Some of the common concerns include the following:

- **Reducing barriers to trade will reinforce the tendency for countries to export commodities that make use of resource-intensive production factors.** As a result of weak environmental policies, trade liberalization in developing countries may result in shifts in the composition of production, exports, and foreign direct investment (FDI) to more pollution- or resource-intensive sectors.
- **Trade liberalization may directly affect environmental standards.** Intensified competition could lead to a "race to the bottom" as governments lower standards in the hope of giving domestic firms a competitive edge in world markets or attracting foreign investment.
- **"Environmental tariffs" may be employed against trading partners deemed to have inadequate environmental standards.** The risks associated with these tariffs are that they may be disguised protection of domestic firms.

In practice, however, the opposite often seems to be the case: most countries that are more open to trade adopt cleaner technologies more quickly, and increased real income is often associated with increased demand for environmental quality (WTO 2004). Greater openness to trade also encourages cleaner manufacturing, because protectionist countries tend to shelter pollution-intensive heavy industries. However, it is often the case that pressures on natural resources, including incentives to overexploit or deplete resources, are generally more directly related to policies and institutions within the sector than to trade openness per se (World Bank 1999).

Some more recent studies have looked at the interface of trade and political economy issues and their implications for the environment and natural resources (Barbier, Damania, and Lèonard 2005; Fredriksson and Mani 2004; Fredriksson, Vollebergh, and Dijkgraaf 2004). These studies highlight the role of lobbying groups in influencing both trade and environmental policy outcomes.

- Is there leakage of energy-intensive industries from OECD countries to developing countries on account of the prescriptions' impact on industries' competitiveness?
- Under what conditions can one justify trade measures under the WTO regime? What are the impacts of levying trade measures on trade flows and emissions?
- What are the underlying trade and investment barriers to the use of clean energy technologies in developing countries?
- In addition to tariff and nontariff barriers, do other issues affect the diffusion of clean energy technologies in developing countries?
- Is liberalization of renewable and clean coal technologies a plausible solution to helping developing countries achieve a low-carbon growth path?
- The Doha Round of negotiations on environmental goods and services provides an opportunity for addressing clean technology transfer issues over the business-as-usual scenario. What conditions are necessary for negotiating a climate-friendly package under the current WTO framework?

The broad objective of this study is to analyze areas in which the climate change agenda intersects with multilateral trade obligations. The study identifies the key issues at stake, as well as possible actions—at the national and multilateral levels—that could help developing countries strengthen their capacities to respond to emerging conflicts between international trade and global climate regimes while taking advantage of new opportunities. The study also attempts to respond to the need for more sector-specific analysis.

Chapter 2 contributes to the literature by exploring the economic, environmental, and political rationale underlying the potential tension between implementation of the Kyoto Protocol and the existing WTO principles. The chapter further identifies areas where priorities for proactive policy initiatives could minimize potential damage to both trade and global environmental regimes. Chapter 3 explores and identifies key barriers and opportunities to spur the transfer and diffusion of climate-friendly and clean-energy technologies in developing countries. It further identifies policies and institutional changes that could lead to the removal of barriers and increased market penetration of climate-friendly technology. Chapter 4 examines and builds on the different approaches that have emerged in the negotiations surrounding trade in environmental goods and services, and it proposes a framework for integrating climate objectives in the discussions. Chapter 5 presents our conclusions and provides a framework for integrating and streamlining the global environment within the global trading system.

Findings and Recommendations

In an attempt to advance the trade and climate change agendas, this report presents the following key findings and recommendations.

Findings

Industrial competitiveness in Kyoto Protocol–implementing countries suffers more from energy efficiency standards than from carbon taxation policies. Though the Kyoto Protocol didn't come into force until 2005, in the 1990s most OECD countries had already established regulatory and fiscal policies, emissions trading systems, and voluntary agreements to combat GHG emissions. Efforts by countries to reduce emissions to meet and exceed Kyoto targets have raised issues of competitiveness in countries that are implementing these policies. The analysis in chapter 2 suggests that efficiency standards are more likely to adversely affect industrial competitiveness than carbon taxes. Some industries—such as metal products and transport equipment—are more severely affected by the increasing efficiency requirements. For those industries, the analysis also suggests that it does not matter whether such standard requirements are imposed by the exporting country, the importing country, or both.

The effects of carbon taxation policies on industrial competitiveness are often offset by "policy packages." Though competitiveness issues have been much debated in the context of carbon taxation policies, the study finds no evidence that industries' competitiveness is affected by carbon taxes. In fact, the analysis suggests that exports of most energy-intensive industries increase when a carbon tax is imposed by the exporting countries, or by both importing and exporting countries. This finding gives credence to the initial assumption that recycling the taxes back to the energy-intensive industries by means of subsidies and exemptions may be overcompensating for the disadvantage to those industries. A closer examination of specific energy-intensive industries in OECD countries shows that only in the case of the cement industry has the imposition of a carbon tax by the exporting country adversely affected trade. In the case of the paper industry, trade actually increases as a result of a carbon tax. Results also suggest that trade is not affected when both countries impose the tax.

Some evidence supports relocation (leakage) of carbon-intensive industries to developing countries. A gradual increase in the import-export ratio of energy-intensive industries in developed countries—and a gradual decline in the ratio in some developing regions—indicates that energy-intensive production is gradually shifting to developing countries as a result of many different factors, including climate change measures in developed countries. Although the trend is converging, the import-export ratio is still greater than 1 in developing countries and less than 1 for developed countries, suggesting that developing countries continue to be net importers of energy-intensive products. Lack of strong evidence of relocation suggests that while the overarching objective of climate policies is to reduce emissions, these policies have been designed to shield the competitive sectors of industrialized economies. More stringent climate policies in industrialized countries in the future may continue to provide the necessary impetus for a more visible leakage of carbon-intensive industries.

Trade measures can be justified only under certain conditions. If a country adopts a border tax measure or even resorts to an outright import ban on products from countries that do not have carbon restrictions, such measures could be in violation of the WTO rules unless they can be justified under the relevant GATT rules. Articles XX(b) and (g) allow WTO members to justify GATT-inconsistent measures, either if these are necessary to protect human, animal, or plant life or health, or if the measures relate to the conservation of exhaustible natural resources, respectively. However, Article XX requires that these measures not arbitrarily or unjustifiably discriminate between countries where the same conditions prevail, nor constitute a disguised barrier to trade. Since most climate change measures do not directly target any particular products, but rather focus on the method by which greenhouse gases may be implicated related to production, issues related to process and production methods (PPMs) are critical for the compatibility between the WTO and Kyoto regimes. In the recent Shrimp-Turtle dispute,[6] the WTO Dispute Settlement Panel and the Appellate Body may have opened the doors to the permissibility of trade measures based on PPMs.

The proposed EU "Kyoto tariff" may hurt the United States' trade balance. There is increasing industry pressure in the EU to sanction U.S. exports for not adhering to the Kyoto targets. This has resulted in calls for a Kyoto tariff on a range of U.S. products to compensate for the loss in competitiveness. Simulation analysis undertaken for this study finds that the potential impact of such punitive measures by the EU could result in a loss of about 7 percent in U.S. exports to the EU. The energy-intensive industries such as steel and cement, which are the most likely to be subject to these provisions and thus would be most affected, could suffer up to a 30 percent loss. Actually, these are conservative estimates, given that they do not account for trade diversion effects that could result from the EU shifting to other trading partners whose tariffs could become much lower than the tariffs on the United States.

Varied levels of tariff and nontariff barriers (NTBs) are impediments to the diffusion of clean energy technologies in developing countries. While the current Kyoto commitments for GHG emissions reduction apply only to Annex I countries, the rising share of developing-country emissions resulting from fossil fuel combustion will require future commitment and participation of developing countries, particularly large emitters like China and India. Some developing countries have already taken measures to unilaterally mitigate climate change; for instance, they have increased expenditures on R&D for energy efficiency and renewable energy programs. It is important that these countries identify cost-effective policies and mitigation technologies that contribute to long-term low-carbon growth paths. Especially for coal-driven economies like China and India, investments are critical in clean coal technology and renewable energy such as solar and wind power generation. Detailed analysis undertaken for the study in chapter 3 suggests that varied levels of tariffs and NTBs are a huge impediment to the transfer of these technologies to developing countries. For example, energy-efficient lighting in India is subject to a tariff of 30 percent and a nontariff barrier equivalent of 106 percent.

Recommendations

A closer examination of the "policy bundle" or package associated with energy taxation is warranted. The results emerging from the analysis in chapter 2 suggest that carbon taxation policies do not adversely affect the competitiveness of energy-intensive industries. This finding suggests that complementary policies (implicit subsidies, exemptions, etc.)—which are used in conjunction with carbon taxation policies levied by Kyoto Protocol–implementing countries, particularly on energy-intensive industries—could be negating any impact of carbon taxation. A more detailed study of this issue is warranted, as it will yield a greater understanding of the implicit subsidies or costs that are associated with each industry. The importance of this finding cannot be understated, as trade measures are justified based on perceptions of higher costs for energy-intensive industries in developed countries and associated loss of competitiveness on account of those costs. The political economy of carbon taxation policies may be used to gain greater insights into the policy package as well.

It would be useful at the outset for trade and climate regimes to focus on a few areas where short-term synergies could be exploited. The energy efficiency and renewable energy technologies needed to meet future energy demand and reduce GHG emissions below current levels are largely available. The WTO parties can do their part by seriously considering liberalizing trade in climate-friendly and energy-efficient goods as a part of the ongoing Doha negotiations to support Kyoto. Within the UNFCCC, it would also help to accelerate and bring greater clarity to the technology transfer agenda. Within the Kyoto Protocol, the most important priority regarding the linkage to trade would be to facilitate a uniform approach to the pricing of greenhouse gas emissions.

Removal of tariff and nontariff barriers can increase the diffusion of clean technologies in developing countries. As stated above, access to climate-friendly clean energy technologies is especially important for the fast-growing developing economies. Within the context of the current global trade regime, the study finds that a removal of tariffs and NTBs for four basic clean energy technologies (wind, solar, clean coal, and efficient lighting) in 18 of the high-GHG-emitting developing countries will result in trade gains of up to 13 percent. If translated into emissions reductions, these gains suggest that—even within a small subset of clean energy technologies and for a select group of countries—the impact of trade liberalization could be reasonably substantial.

Streamlining of intellectual property rights, investment rules, and other domestic policies will aid in widespread assimilation of clean technologies in developing countries. Firms sometimes avoid tariffs by undertaking foreign direct investment (FDI) either through a foreign establishment or through projects involving joint ventures with local partners. While FDI is the most important means of transferring technology, weak intellectual property rights (IPR) regimes (or regimes perceived as weak) in developing countries often inhibit diffusion of specific technologies beyond the project level. Developed country firms, which are subject domestically to much stronger IPRs, often transfer little knowledge along with the product, thus impeding widespread dissemination of the much-needed technologies. Further, FDI is also subject to a host of local country investment regulations and restrictions. Most non–Annex I countries also have low environmental standards, low pollution charges, and weak environmental regulatory policies. These are other hindrances to acquisition of sophisticated clean energy technologies.

The huge potential for trade between developing countries (South-South trade) in promoting clean energy technology in those countries needs to be explored more. Traditionally, developing countries have been importers of clean technologies, while developed countries have been exporters of clean technologies. However, as a result of their improving investment climate and huge consumer base, developing countries are increasingly becoming major players in the manufacture of clean technologies. A key development in the global wind power market is the emergence of China as a significant player, both in manufacturing and in investing in additional wind power capacity. Similarly, other developing countries have emerged as manufacturers of renewable energy technologies. India's photovoltaic (PV) capacity has increased several times in the last four years, while Brazil continues to be a world leader in the production of biofuels. These developments augur well for a buoyant South-South technology transfer in the future.

Clean technology trade would greatly benefit from a systematic alignment of harmonization standards. The volume of trade and the level of tariffs can be examined by identifying and tracking the unique HS code associated with each technology or product under the Harmonized Commodity Description and Coding System (commonly called the harmonized system or HS). Typically, each component of the technology has a different HS code. At the WTO-recognized six-digit code level, clean energy technologies and components are often found lumped together with other technologies that may not necessarily be classified as being beneficial to either the global or even local environment. Solar photovoltaic panels are categorized as "Other" under the subclassification for light-emitting diodes (LEDs). Such categorization suggests that reducing the customs tariff on solar panels might also result in tariff reduction for unrelated LEDs. Similarly, clean coal technologies and components are not classified under a separate category, and all gasification technologies are lumped together. The imprecise definition also raises another issue for countries that are considering removal of trade barriers to clean energy equipment and components. In cases where the codes are not detailed enough, the scope of the tariff reduction may become much broader than anticipated.

The ongoing WTO negotiations on environmental goods have the potential to contribute significantly to both trade and climate change efforts, but the negotiations will need to address a number of challenges. Liberalizing trade in specific goods and technologies that are relevant for climate change mitigation may have implications with regard to the costs of mitigation measures, particularly those technologies that face high tariff and nontariff barriers to trade. The relevant concerns cannot be disregarded, such as those related to definition of relevant products (especially products that also have nonenvironmental uses); harmonizing classifications and descriptions across countries within the harmonized system; changes in technology; issues related to perceived impacts on domestic industries; and nontariff measures and access to technology. Goods that would benefit include those that directly address climate change mitigation, as well as environmentally preferable products that contribute to zero or reduced GHG emissions during production, consumption, or use. Goods and technologies used in CDM projects (including programmatic CDMs) are particularly relevant.

Political economy dynamics may necessitate the consideration of innovative packages for trade liberalization in climate-friendly goods. One package could be an ITA-type agreement within single undertaking, whereby members representing a minimum percentage of trade in climate-friendly products would join. Such an agreement could be a subcategory within any larger negotiated package of environmental goods or in a separate agreement. A second option, particularly if negotiations on environmental goods fail to reach a meaningful outcome, would be to consider a plurilateral agreement similar to the agreement on government procurement. In that option, the agreement could come into effect immediately or even independent of the conclusions of the Doha Round negotiations, but only the signatories would extend as well as receive the benefits of trade liberalization in climate-friendly products. The advantage in the second option would be that members, particularly developing countries, need not feel compelled to sign on immediately.

RTAs also offer opportunities, but there are challenges to consider. A collapse of the Doha Round could result in a spurt in regional trade agreements (RTAs) as more WTO members seek alternative routes to pursue their trade agenda. A number of problems associated with defining environmental and climate-friendly goods will be less of an issue, as most RTAs would normally liberalize at a broader HS level (usually six-digit). With regard to provisions aimed at building supply-side capacities and technical assistance, RTAs may be better suited to include provisions tailored to the needs of participating developing countries. On the other hand, RTAs may also result in the diversion of trade from countries that are most effective at producing climate-friendly technologies if those countries are excluded from an RTA.

Making tangible and immediate progress is necessary in several venues. Just as business as usual in GHG emissions is not sustainable, business as usual in trade negotiations is not an adequate response to challenges posed in the study. At least some of the steps mentioned could be taken in the context of the Doha Round and perhaps even agreed to separately if WTO members fail to come to an agreement and the Doha Round is terminated or suspended indefinitely. Although the role of WTO negotiations has been emphasized in this study, there are other venues where similar progress can be made. In particular, the next COP/MOP (Conference/Meeting of the Parties to the Protocol) meetings in 2007 and the G-8+5 summit in 2008 both offer opportunities for the leaders of the major GHG-emitting countries to make specific commitments to reduce tariff and nontariff barriers to international trade and investment in goods, services, and technologies that contribute to the mitigation of climate change.

Notes

1 Competitiveness concerns were the explicit prime motivation for the withdrawal of the United States from the Kyoto process. Competitiveness concerns have since plagued Canada, the United States' largest trading partner and the bearer of a relatively difficult emissions reduction target.

2 The Montreal Protocol on Substances that Deplete the Ozone Layer is one of the first international environmental agreements to include trade sanctions to achieve the stated goals of a treaty. It also offers major incentives for nonsignatory nations to sign the agreement. The treaty negotiators justified the sanctions because depletion of the ozone layer is an environmental problem most effectively addressed on the global level. Furthermore, it was argued that without the trade sanctions, there would be economic incentives for nonsignatories to increase production, damaging the competitiveness of the industries in the signatory nations as well as decreasing the search for less-damaging CFC alternatives. Article IV of the Montreal Protocol stipulated that one year after the treaty came into force, all imports of controlled substances "from any non-party states are banned and that none of the signatories are allowed to export a controlled substance to non-party states."

3 The UNFCCC uses the term *environmentally sound technologies* for climate-friendly technologies. This paper uses the term *clean energy technologies* to be consistent with the Clean Energy Investment Framework (CEIF).

4 The issue of trade and the environment has surfaced at the World Bank from time to time. Two edited volumes (World Bank 1992, 1999) focused on issues such as pollution havens, "race to the bottom," and foreign direct investment inflows. These were quite useful in informing the broader discussion in the area at that time.

5 The traditional arguments of trade and growth, which are often positively associated with local pollution issues, do not in fact hold for global externalities such as greenhouse gas emissions. This is due to the classic "free rider" problem. Any country individually would have little incentive to cut back emissions, because it would bear the costs alone even though the benefits would accrue to all.

6 *United States—Import Prohibition of Certain Shrimp and Shrimp Products*, WT/DS58/AB/R. See chapter 2.

CHAPTER 2

Climate Change Policies and International Trade: Challenges and Opportunities

ALTHOUGH THE KYOTO PROTOCOL to the UNFCCC came into force only in 2005, a number of OECD countries had policies and other measures in place to combat greenhouse gas emissions, even going back to the 1990s. Nevertheless, efforts to reduce emissions to meet and exceed Kyoto targets have raised issues of competitiveness in countries that are implementing these policies, as well as fear of leakage of carbon-intensive industries to nonimplementing countries. This has also led to proposals for tariff or border tax adjustments to offset any adverse impact of capping CO_2 emissions.

In this chapter, we consider the following: (i) the implications of climate change policies on competitiveness across industries, as well as issues related to leakage, if any, of carbon-intensive industries to developing countries; (ii) both theoretical and practical implications of the proposed tariff or border tax measures, including their compatibility with existing WTO rules; and (iii) possible synergies between the Kyoto and WTO regimes.

Do Climate Change Measures Affect Competitiveness?

There is a widespread concern regarding international competitiveness of major industries, especially in the energy-intensive sector, among countries that have undertaken several measures to reduce GHG emissions. These countries especially worry that higher energy costs not only burden them domestically but also give competitors in countries that do not have these measures (especially the United States and China) a competitive edge and an unfair advantage.

Generally, climate change measures can be grouped as regulatory measures, fiscal measures, market-based instruments, or voluntary agreements (see appendix 2 for a detailed description of each specific measure). As illustrated in table 2.1, the choice of policy instruments differs significantly across nations, reflecting institutional, economic, and policy structures. The higher costs usually accrue from fiscal and regulatory measures, or a combination of these measures, that are levied by these countries.

This section analyzes the impacts of GHG-emissions-reducing measures on the export competitiveness of energy-intensive sectors in OECD countries.

Scope and Analytical Framework

This study focuses on two types of instruments: (i) carbon taxes associated with a fiscal measure, and (ii) energy efficiency standards associated with a regulatory measure. The reason for choosing them is that both have been in existence for quite some time in many countries; hence, the impacts on competitiveness are much more traceable compared to the emissions trading and voluntary regimes, which are more recent. While both carbon taxes and energy efficiency standards aim to reduce energy consumption, as discussed below, they use very different mechanisms to reduce emissions.

Carbon tax. A carbon tax is a tax on the carbon content of fuels (principally coal, oil, and natural gas) that generate CO_2 emissions when burned. The tax would apply at a specific rate per ton of coal, per barrel of oil, or per million cubic feet of gas, with the amounts adjusted to equalize implied taxes on carbon content.[1] The rationale of such a tax is to reduce GHG emissions primarily responsible for climate change.[2] Since private sector decisions do not take adequate account of their wider effects, a tax can serve to correct what would otherwise be socially excessive emissions. Carbon tax measures used here also consider broader energy-input taxes used in some countries (see table 2.2 for the status of carbon tax measures in selected OECD countries).

Effect of a carbon tax (or a similar energy input tax) on competitiveness. A carbon tax would affect competitiveness by increasing the costs of polluting inputs (e.g., coal, oil, natural gas, and electricity). Hence, a carbon tax may significantly increase production costs, leading to lower profits, either through lower margins or through a reduction in sales (or both). A tax may not necessarily lead to a one-for-one reduction of profit margins. Part of the tax may be borne by input suppliers and part by the final consumers.

The impact of a carbon tax would also differ across the sectors of the economy because of different input combinations and emission profiles. A recent OECD study (2006) identified three factors driving sectoral competitiveness resulting

TABLE 2.1
Existing Measures to Combat Climate Change in Annex I Countries

Country/Region	Measure
Regulatory Measures	
Regulatory instruments (regulations, standards, directives, and mandates) have been most commonly used to promote energy efficiency and renewable energy, including cogeneration and low-emission motor vehicles in OECD countries.	
EU	The EU Renewable Electricity Directive of 2001 seeks to increase the share of renewable energy production to 12 percent and renewable electricity generation to 22 percent. A 2004 directive on combined heat and power (CHP; Directive 2004/8/EC) provides a framework for promoting and developing high-efficiency cogeneration. The EU's Energy Performance of Buildings directive (Directive 2002/91/EC) requires member states to adopt energy performance standards and has introduced energy labeling of buildings. Under the EU's directive on energy labeling of domestic household appliances (Directive 1996/75/EC), domestic household appliances sold in the EU must carry a label grading them according to their energy efficiency.
U.K., Austria, Belgium, Italy, Netherlands, and Sweden	The 2001 Renewables Obligation requires suppliers to use renewable sources for a specific and annually increasing percentage of the electricity they supply, to meet a target of 10 percent of electricity from renewable sources by 2010.
U.K.	The U.K. government set a new target to achieve at least 10,000 MWe of installed "Good Quality CHP" capacity by 2010.
Sweden	The Environmental Code in Sweden (1999) stipulates that the best possible technology should be used in all industrial operations.
Japan	In force since April 1999, the revised Energy Conservation Law sets energy conservation standards for home/office appliances and fuel efficiency standards for autos.
Canada	Canada has recently pursued a strategy to cut greenhouse gas emissions per unit of production by 18 percent by 2010 by setting mandatory reduction targets for major industries.
Fiscal Measures	
Considered as one of the most effective instruments for environmental objectives, fiscal measures usually include carbon/energy taxes that are based on the carbon or energy content of the energy products.	
Finland	Finland introduced a carbon tax in 1990, based on the CO_2 content of the fuel, starting at a comparatively low level of Mk 6.7 per ton of CO_2 (US$1.2/t CO_2).

(continued)

TABLE 2.1
Existing Measures to Combat Climate Change in Annex I Countries *(Continued)*

Country/Region	Measure
Sweden	As part of an overall fiscal reform, Sweden introduced a carbon tax and a value added tax on energy, and lowered the existing energy tax.
Norway	Norwegian authorities introduced carbon taxes in 1991 with a tax rate that differed across fossil fuel categories and the geographic location of the activity.
EU	The EU negotiated a minimum tax directive concerning energy products and electricity (Directive 2003/96/EC); the directive entered into force in the beginning of 2004.
Market-Based Instruments	
These instruments are based on the premise that "free markets find optimal solutions." They include emissions trading and tradable renewable energy certificates (TRCs) as effective means to help decrease the cost of mitigating greenhouse gas emissions.	
U.K.	U.K. Emissions Trading Scheme (ETS) is the first economy-wide greenhouse gas emissions trading scheme.
EU	EU ETS is the largest company-level trading system for CO_2 emissions in terms of its value and volume.
Japan	Launched in 2005, the Japanese voluntary emission trading scheme seeks to implement measures to promote cost-efficient emissions reductions and to accumulate knowledge and experience in domestic ETS.
Voluntary Agreements (VAs)	
Voluntary agreements differ from other measures in that they are negotiated directly between governments and industry/firms rather than result from mandates imposed by the governments; they are often the preferred policy approach from industries' perspective.	
Japan	Japan's voluntary action plan, "Wisdom of Industry," covers 82 percent of CO_2 emissions from industry/energy conversion sectors (34 subsectors) and is expected to deliver about 30 percent of the needed energy savings and the related emission savings.
Netherlands	VAs in the Netherlands, in combination with fiscal incentives and environmental permits, are the main policy tool used to limit industry GHG emissions.
EU	Voluntary commitments by European, Japanese, and Korean carmakers would reduce CO_2 emissions from cars sold in the EU by 25 percent by 2008–09.

TABLE 2.2
Status of Carbon Tax Regimes in Selected OECD Countries

Country	Status	Tax Type
Australia	Proposed in 1994, not adopted	Greenhouse levy
Austria	2000 (updated)	Energy tax
Belgium	Planned	Energy tax
Denmark	1993 (implemented), 1996 (updated)	Carbon tax (part of a tax reform)
Estonia	2000 (implemented)	Carbon tax
EU	Proposed since 1991 but lacks support from some members	CO_2/environment tax
Finland	1990 (implemented), 1998 (updated)	Carbon/energy tax
France	1999 (proposed), 2000 (suspended)	Energy/carbon tax
Germany	1999 (implemented)	Energy tax (ecotax)
Italy	1998 (implemented), 1999 (revised), then suspended	Energy tax reform
Japan	Pending	Carbon tax
New Zealand	2007 (planned)	Carbon tax
Norway	1991 (implemented), 1999 (updated)	Carbon tax
Poland	Pending	Carbon tax
Portugal	Pending	Carbon tax
Slovenia	1997 (introduced)	Carbon tax
Sweden	1991 (implemented), 2001 (updated)	Carbon tax (part of a tax reform)
Switzerland	Pending	Carbon tax
Netherlands	1996 (implemented)	Energy tax
United Kingdom	2001 (implemented)	Climate change levy
United States	Proposed in 1993, not adopted	BTU tax

Sources: IEA, OECD, EEA (various years). See appendix 2 for details of various measures.

from an environmentally related tax. According to the OECD study, the effects on competitiveness will be stronger under the following conditions:

- *The lower the ability to pass on cost increases in prices.* International competition is the most important factor in reducing the ability to pass on cost increases, followed by the price responsiveness of demand, and the market structure and the geography of the sector market.
- *The lower the feasibility of the substitution possibilities,* because limited scope for identifying and financing cleaner production technologies and processes implies an inability to substitute away from environmental taxes.
- *The higher the energy intensity of the sector*, since the bulk of the tax is levied on energy use and transportation.

In a country that imposes a carbon tax (or a similar energy input tax), the expectation is that energy-intensive industries will likely suffer from a significant increase in production costs compared with their trading partners. Consequently, these industries either will become less competitive internationally and lose some of their market share or, in order to avoid this loss, will migrate to countries with no such taxes. In each case, exports of energy-intensive commodities with the carbon tax will decrease, while their imports will likely increase. Conversely, a carbon tax imposed by an importing country will make its import-competing industries less competitive, thereby benefiting countries exporting to this country.

In anticipation of the adverse terms of trade affecting their most competitive sectors, many countries provide either a full or partial exemption for energy-intensive industries and export industries. In many cases, energy products used mainly by heavy industries are exempted from tax. Most countries do not tax coal at all, while a few countries that have taxes on these products grant very significant exemptions (OECD 2006). In other cases (Denmark, Germany, Sweden, and the United Kingdom), reduced tax rates combined with generous rebates are applied to industry with respect to carbon or other energy taxes (see appendix 2 for some standard exemptions given for a carbon tax). These considerations often make it difficult to measure competitiveness impacts at the national level.

Countries that levy domestic taxes on fossil fuels for fiscal purposes (e.g., excise tax) apply a border tax adjustment equal to the domestic tax when importing such fuels. However, no such border tax adjustment schemes exist in practice for energy inputs used in the production of final goods (Biermann and Brohm 2003). Hence, competitiveness related to efforts to significantly reduce GHG emissions continues to be a major point of debate, especially in terms of the negative impacts on the international competitiveness of some energy-intensive sectors. This debate has derailed any efforts in the United States to impose a carbon tax, or in the EU to institute a common framework on energy taxation.

Energy efficiency standards. Energy efficiency standards and labeling schemes for appliances and equipment now play an important role in many OECD countries' energy and environmental strategies. Energy efficiency standards may be designed and implemented in many different ways, for example, as technical specifications or as industry norms implemented through regulations or voluntary agreements. Performance standards for electrical appliances, usually known as minimum energy performance standards (MEPS), are now common and impose a minimum energy efficiency rating or a maximum consumption rating for all the products on the market.

Efficiency standards set levels in a number of different ways. In Europe, a statistical approach is used. The energy efficiency of appliances already on the market is used as a basis and the standard is drawn up to obtain an improvement of 10 to 15 percent in the average energy efficiency of new appliances. In other countries, regulations are based on a cost-benefit evaluation (e.g., in the United States, to raise the energy efficiency of appliances to a level that corresponds to a three-year return on investment).

Several EU countries introduced voluntary agreements in the 1980s and 1990s (Germany in the 1980s, Nordic countries in the 1990s, Switzerland in 1995). Since 1999, an EU directive has defined mandatory energy efficiency standards for refrigerators and freezers in EU countries. Japan continues to have a voluntary target for energy efficiency improvement by a given year (table 2.3).

The cost and time needed to comply with different energy efficiency program requirements could add to the cost of internationally traded products. However, since regulations could, in principle, be applied equally to imports and locally manufactured products, effects on trade in countries with higher MEPS could be nullified to some extent. On the other hand, standards could adversely affect trade from countries with lower or no standards to countries that have higher efficiency standards.

Empirical Specification

In this study we used a standard gravity model of trade to gauge the effects of these two measures—carbon taxes and energy efficiency standards—on OECD countries' exports. The basic gravity model—as developed by Tinbergen (1962) and Linnemann (1966)—predicts bilateral trade flows based on the economic sizes of (often using GDP measurements) and distance between two units. Some models include, alongside distance, the land areas of the trading partners (proxy for transport cost within the country), tariff and price variables, as well as a variety of proxies for "closeness" between the trading partners, such as contiguity, common language (cultural affinity), and trading bloc membership. This model is often used to examine bilateral trade patterns in search of evidence on "natural" (noninstitutional) regional trading blocs, to estimate trade creation and trade diversion effects from regional integration, and to estimate trade potential for new entrants to a trading bloc.

The gravity model can also be augmented by variables that measure strictness of environmental regulations, both in the importing and exporting countries (Harris, Kónya, and Mátyás 2002; van Beers and van den Bergh 1997). An advantage of using a bilateral trade model rather than a multilateral trade model is that the

TABLE 2.3
Existing Energy Efficiency Standards for Select Products in OECD Countries

	Mandatory	Voluntary
Refrigerators	EU, Norway, Hungary, Canada, Korea (Rep. of), Mexico, New Zealand, United States	Switzerland, Japan, United States
Washing machines	EU, Norway, Hungary, United States	United States
Air conditioning	Canada, Korea (Rep. of), Mexico, United States	Japan, United States
Lamps	EU, Norway	United States

Note: In the United States, mandatory or voluntary depends on states.

effects on trade flows between countries as a result of differences in strictness of environmental regulations may cancel out in multilateral models where trade is an aggregate of bilateral trade flows.

The model uses the industry-level bilateral exports between two countries relative to the product of the two countries' GDPs as a dependent variable. The explanatory variables include distance between the two countries, variables that proxy common borders, common currency, and common free trade agreements (see appendix 3 for detailed model specification and results). To understand the separate impacts of carbon taxes and energy efficiency standards, we introduce separately two sets of additional variables to capture the effects on exports relative to the baseline scenario when no such taxes or standards are in place. The variables are designed to capture a scenario where only an exporting country has a carbon tax (or energy efficiency standards) in the year; the second scenario, where only an importing country has a carbon tax (or energy efficiency standards) in the year; and the third scenario, where both countries have carbon taxes (or energy efficiency standards) in the year. The expected results are summarized in table 2.4.

How these two policies affect specific industries is another issue that requires adequate consideration. For that reason, the study also assesses the effects of these two instruments on energy-intensive industries (namely, paper and paper products, industrial chemicals, nonmetallic products, iron and steel, and nonferrous metal) and industries that produce outputs subject to higher energy efficiency standards (namely metal products, machinery, electrical machinery, transport equipment, and scientific equipment).

Data

The study uses a panel of industry data from the OECD countries spanning 1988 to 2005. The main data source is the WITS (World Integrated Trade Solution)

TABLE 2.4
Predicted Competitiveness Impacts of Carbon Taxes and Energy Efficiency Standards

Carbon Tax	Impact on the Exporting Country	Energy Efficiency Standard	Impact on the Exporting Country
Carbon tax by an exporting country	Negative	Energy efficiency standards in the exporting country	Neutral or marginally negative
Carbon tax by an importing country	Positive	Energy efficiency standards in the importing country	Negative
Both exporting and importing countries have carbon tax	Neutral or marginal decline in trade	Energy efficiency standards in both exporting and importing country	Neutral or marginal decline in trade

database which provides the value of exports at the three-digit ISIC (International Standard of Industrial Classification) level for all OECD countries. GDP figures were obtained from the World Development Indicators (World Bank 2006b). The gravity variables, such as bilateral distance between country pairs, and the common border variable are from Nicita and Olarreaga (2004). Information on carbon taxes and energy efficiency standards were obtained from various national sources, as described in appendix 2.

Caveats

First, a limitation of this analysis is that it uses climate change measures, namely carbon taxes and energy efficiency standards, as binary variables—1 if a country has carbon taxes (energy efficiency standards) and 0 otherwise. The variables do not reflect the differentiated levels of standards and taxes that are levied in different countries and across the different fuels. Thus, results need to be interpreted with some degree of caution, as the analysis is unable to provide a direct assessment of the extent of trade loss or gain from the levels of stringency across countries. Nonetheless, by comparing countries with and without measures, we gain useful insights to the dynamics of climate change measures on country competitiveness. It is this issue that has dominated the debates, not the actual levels.

Second, carbon tax values or energy efficiency standards could change with time even for a given country. However, data constraints prevent a more detailed examination of this phenomenon.

Results

From the analysis, we find that both carbon taxes and energy efficiency standards have a statistically significant negative effect on competitiveness through their impacts on bilateral trade flows (depending on the specifications imposed in the modeling). This is particularly true when the focus is on industries that are subject to higher energy efficiency standards and are not subsidized by governments. This adverse effect is missing when the focus is on energy-intensive industries that usually receive government subsidies. Appendix 3 presents the detailed regression results of the various model specifications. The results are summarized below in table 2.5, which pools all manufacturing industries for all the OECD countries in all the sampled years.

The regressions first examined the impact of only a carbon tax. Results show that export competitiveness is adversely affected only when importing countries impose a carbon tax. A carbon tax imposed by exporting countries does not seem to matter. This could be because most countries that have a carbon tax also actively subsidize or exempt those energy-intensive industries (from a carbon tax), which also happen to be in their competitive sectors.

TABLE 2.5

Impact of Carbon Taxes and Energy Efficiency Standards on Export Competitiveness

	Carbon Tax (imposed by country)			Energy Efficiency Standards (imposed by country)		
Measures	Exporting	Importing	Exporting and Importing	Exporting	Importing	Exporting and Importing
Carbon tax only		Marginally significant (−)				
Energy efficiency standards only				Highly significant (−)	Highly significant (−)	Highly significant (−)
Carbon taxes and energy efficiency standards		Marginally significant (−)		Highly significant (−)	Highly significant (−)	Highly significant (−)
Energy-intensive industries	Highly significant (+)	Highly significant (−)	Highly significant (+)			
Industries subject to energy efficiency standards				Highly significant (−)	Highly significant (−)	Highly significant (−)

(−) denotes a decrease in trade and (+) denotes an increase in trade.

The regressions then examined the impact on trade flows by considering only the effects of energy efficiency standards. Strong negative effects on export competitiveness are found, irrespective of whether the standard is imposed by exporting countries, importing countries, or both. Bilateral trade, on average, decreases by nearly 10 percent in all cases. When both carbon taxes and energy efficiency standards are included in the model, similar results are obtained. This suggests that these two policies do not interfere with each other when it comes to affecting export competitiveness.

The results in table 2.5 also show that when a carbon tax is imposed only by the importing countries, it adversely affects the competitiveness of exporting countries. This effect could be due to the offsetting measures applied by importing countries to mitigate and nullify the impact of such taxes on domestic industries. On the other hand, when a carbon tax is imposed by the exporting countries, or by both importing and exporting countries, the overall trade between countries increases. This once again suggests that subsidies and other exemptions on those energy-intensive industries may be overcompensating for the disadvantages arising from the imposition of the carbon tax.

We then examined how these policies affect specific industries that use energy intensively. The results, summarized in appendix 4, suggest that the net effect varies considerably across the various industries. Trade competitiveness is adversely affected by a carbon tax in the case of the cement industry, but the paper and steel industries actually benefit from a carbon tax. Similarly, energy efficiency standards mainly affect the transport equipment and metal products industries.

Conclusion

This section provides econometric evidence on the hypothesis that domestic climate change policies affect countries' export competitiveness. The focus was on two policies: (i) carbon taxes, which usually target those industries that use energy intensively, and (ii) energy efficiency standards, which affect those industries whose outputs are usually subject to higher energy efficiency standards. The study finds some evidence of both carbon taxes and energy efficiency standards having negative impacts on trade flows and hence export competitiveness. Evidence on carbon taxes is contrary to the hypothesis when we examine the trade of energy-intensive industries. The subsidies and exemptions for some industries (as documented in appendix 2) are probably so generous that trade actually increases as a result.

In Search of Carbon Leakage: Examining the Relocation of Energy-Intensive Industries to Developing Countries

Many industrialized countries are concerned about the potential impact that mandatory carbon reduction targets would have on their economies. Among these concerns is that any plan that exempts developing countries from emissions limits would not be effective, because carbon-intensive industries would simply shift their operations to one of the exempt countries.

A relocation of carbon-intensive industries, more frequently referred to as "carbon leakage," would not only undercut the environmental benefits of the Kyoto Protocol; in addition, the competitiveness of industrialized-world industries could also suffer. Most emissions in industrialized countries result from inherently domestic activities such as transportation, heating, cooling, lighting, and other such activities, where leakage is either difficult or impossible. On the other hand, for energy-intensive industries such as cement, chemicals, and others, international competitiveness is an important concern. This is somewhat akin to the "pollution havens" debate that dominated the environmental literature in the 1990s.[3]

Within the specific context of the Kyoto Protocol, the IPCC in its 2001 assessment concluded that "the possible relocation of some carbon-intensive industries to non–Annex I [developing] countries and wider impacts on trade flows in response to changing prices may lead to leakage in the order of 5 to 20 percent" (IPCC 2001). Accordingly, in the worst-case scenario, if an emissions reduction of 5 percent were to occur in the industrialized world (roughly what the Kyoto Protocol calls for), 1 out of those 5 percent would not disappear completely, but would instead become developing-world emissions due to shifting industrial activity.

Is Such Leakage Really Happening?

In this section, we examine the evidence for any relocation of carbon-intensive industries due to more stringent climate policies, mostly in the OECD countries. First, we identified industries that will be most affected by carbon reduction targets. As seen earlier, these energy-intensive industries—pulp and paper, industrial chemicals, iron and steel, nonmetallic mineral products, and nonferrous metals—are easily identifiable from the literature (Mani and Wheeler 1998). The analysis begins with the 1990s, when most countries began to implement climate-friendly policies such as the introduction of carbon taxes and energy efficiency standards. We observed global trade trends in these key sectors.

One of the factors influencing the operations of the energy-intensive sectors is the relative energy price in addition to land and labor costs. In energy-intensive sectors, energy costs account for between 10 and 20 percent of the value of sales—not trivial, but also not dominant (Baumert and Kete 2002). In addition, the location decision is also influenced to some extent by domestic market size and growth potential. During the period 1990–2005 that we examined, global energy prices did not experience any out-of-the-ordinary fluctuations except in more recent years (figure 2.1).[4]

During this same period, most developing countries also drastically reduced energy price subsidies, ruling out major price differentials between developed and developing countries. On the other hand, climate-friendly energy policies were being implemented, mainly in many high-income OECD countries, which would

FIGURE 2.1
World Crude Oil Price, 1990-2005

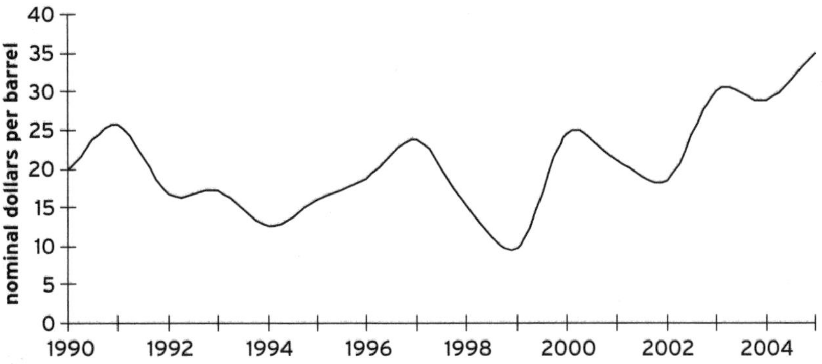

Source: IEA 2006.

entail additional costs on these industries. All else being equal, one would then expect that this would enhance the comparative advantage of low- and middle-income economies in the production of energy-intensive products.

When the actual data are examined on imports and exports across various income groups and regions, this provides some interesting results. The import-export ratio of energy-intensive production in high-income OECD countries shows an increasing trend. When the same ratio is examined for low- and middle-income developing economies, there is almost a mirror image of the OECD graph (figure 2.2). The correlation coefficient between the OECD and low- and middle-income ratios is 0.9. This could be a reflection of some relocation of energy-intensive industries to developing economies, which were not imposing any additional constraints on these industries to mitigate climate change. However, the ratio is still less than 1.0 for OECD countries and more than 1.0 for developing economies, suggesting that OECD countries continue to be net exporters and developing countries are still net importers of energy-intensive products.

The next step is to see if there are any discernable trends within the OECD and developing countries. Given that European countries have been more proactive in implementing climate-friendly policies, we presume they are also experiencing a more pronounced shift in these sectors. As shown in figure 2.3, the United States, and not the EU, has been experiencing much more pronounced movement or leakage of energy-intensive sectors. There could be three possible reasons for this. First, the gradual relocation of energy-intensive industry from the United States could be a way to circumvent any future policy shift in the climate change area. Second, the cause could be other factors such as cheap land, labor, and growing markets in developing countries. Third, the lack of any major shift

FIGURE 2.2
Import-Export Ratio of Energy-Intensive Products in High-Income OECD Countries and Low- and Middle-Income Economies

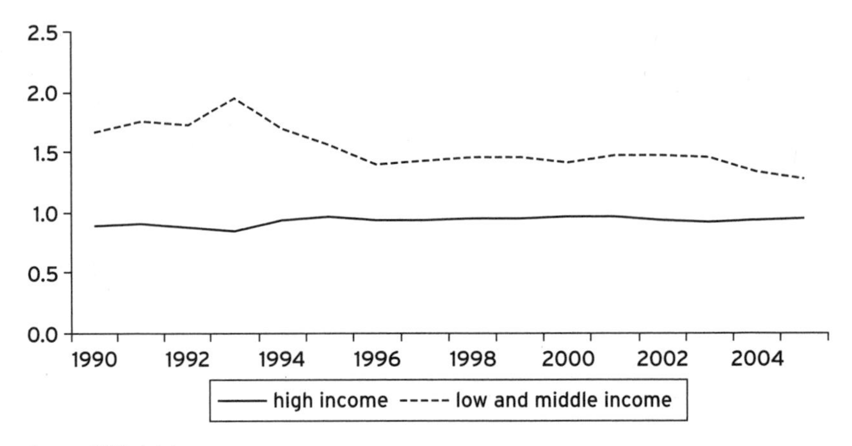

Source: WITS database.

FIGURE 2.3
Import-Export Ratio of Energy-Intensive Products in the United States and EU

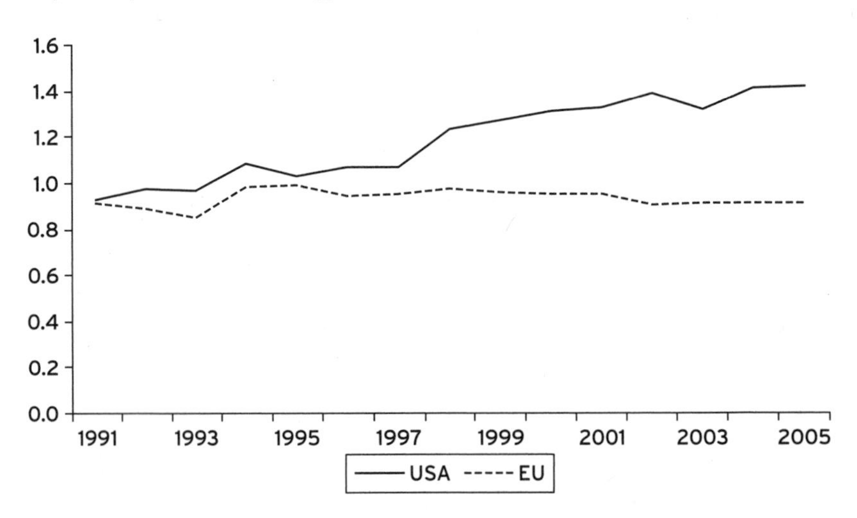

Source: WITS database.

in Europe could be a reflection of the movement of industries within the EU countries, which is not reflected in the aggregates.

The data were further examined to see if some relocation of industries from the United States is mainly to East Asia, and especially China. Though China reflects the general declining trend in import-export ratio observed in East Asia,

FIGURE 2.4
Import-Export Ratio of Energy-Intensive Products in Low- and Middle-Income East Asian and Pacific Economies and China

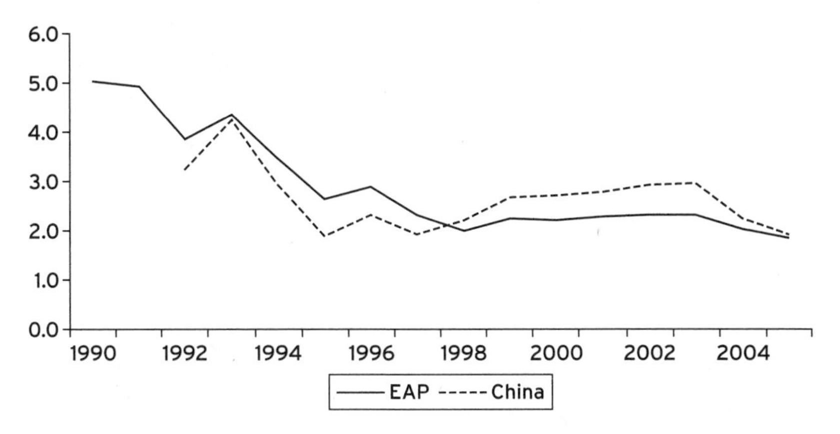

Source: WITS database.

FIGURE 2.5
Import-Export Ratio of Energy-Intensive Products in Low- and Middle-Income Economies in Various Regions

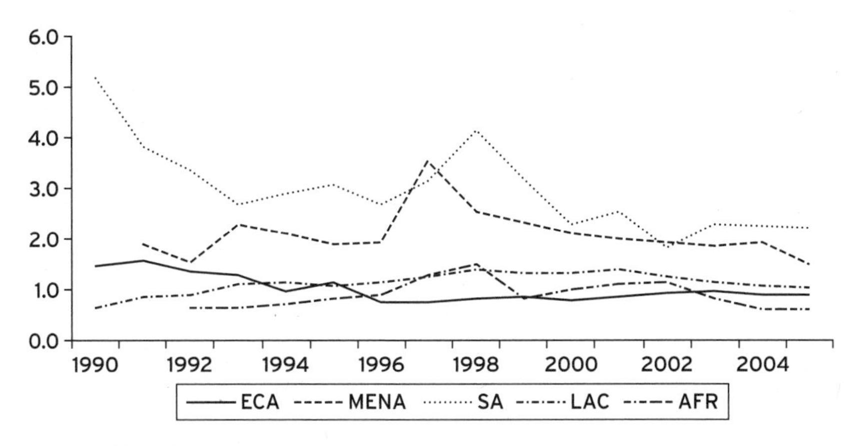

Source: WITS database.
Note: ECA–Europe and Central Asia, MENA–Middle East and North Africa, SAR–South Asia, LAC–Latin America and the Caribbean, AFR–Africa.

it is not driving the trend, because its economic growth probably continues to fuel increased imports of energy-intensive products (figure 2.4).

In terms of other developing regions, there are no discernible trends (figure 2.5). Most of them (except low- and middle-income Europe and Central Asia) seem to have experienced a downward trend toward the late-1990s. To some extent, this could be a reflection of increased energy prices followed by lower imports, more so than a relocation of energy-intensive production from developed countries. In the case of the low- and middle-income economies of Europe, a considerable downward trend is seen in the import-export ratio of energy-intensive products. While some of this trend could be attributed to the general decline of economic activity following transition (and hence lower imports), the decline also may reflect some leakage of carbon- and energy-intensive industries from the United States or EU to take advantage of laxer climate change policies. This relationship could be true, especially given the proximity to EU markets.

Conclusions

This analysis suggests a gradual increase in the import-export ratio of energy-intensive industries in developed countries, and a gradual decline in the ratio in some developing regions. The findings thus suggest some evidence—although not very pronounced—of leakage of carbon- and energy-intensive industries to developing economies that could be attributed to more stringent climate change policies and energy efficiency standards. A detailed econometric analysis would be needed to ascertain the effects more precisely. However, the results do reveal some interesting facts. The ratio is still greater than 1 for developing countries and less than 1 for developed countries, suggesting that developing countries continue to be net importers of energy-intensive products. Among developing regions, East Asia and especially China are emerging as major exporters of energy-intensive products. The convergence of the ratios suggests that in the medium to long run, the increased stringency of climate policies in some industrial countries and increased growth in some developing countries in the next decades could accentuate the existing trends.

Some caveats need to be kept in mind: This analysis is a reflection of climate policies that were put in place long before Kyoto, and whose objective was also to shield the competitive sectors. It is therefore early to analyze the implications of the more recent emissions trading arrangements that have now been put in place. Further, other factors, such as labor market differentials, availability of raw materials, and growing market size of developing economies, could also account for this. Also, even closed economies will tend to have a different composition of production at various stages of development, simply because the composition of domestic demand changes.

Nonetheless, as shown in box 2.1, the recent globalization trends observed in the chemical sector support the evidence presented here. The evidence suggests that increased concentration of energy-intensive sectors in developing countries could also be a signal for those countries' greater future involvement in any post-Kyoto global GHG reduction measures.

BOX 2.1

Globalization of the Chemical Industry

Over the past 30 years, the global chemical industry has experienced steady growth in production, consumption, and trade, with the value of chemical shipments rising from US$171 billion in 1970 to US$1.5 trillion in 1998. OECD countries accounted for 83 percent of world output in 1970, but—despite overall growth at the global level—dropped to 78 percent in 1998 due to stronger growth in non-OECD countries. Industry growth is projected to continue until 2020, but non-OECD countries are expected to experience a greater rate of growth than OECD countries. Over the past 40 years, a global expansion of the chemical industry has occurred owing to the following factors:

- Multinational chemical companies emerged as OECD-based companies that invested in non-OECD countries, a trend that is expected to continue.
- Domestic chemical industries in many developing countries increased investments, began producing specialty chemicals, and increased their exports of bulk chemicals.
- Some countries with a small chemical industry became major suppliers of chemicals, for example, Korea (Rep. of), China, Taiwan (China), Saudi Arabia, and Canada.
- Global markets have developed along with world economic growth.
- There has been a progressive increase in international trade as tariffs and other trade barriers have been reduced.
- Telecommunications and transportation have had significant advances.

Despite the dominant position of the United States, Western Europe, and Japan since the 1970s, other countries initiated or increased their production. For example, in 1975, 65 percent of world production of methanol occurred in developed regions, with 35 percent from the rest of the world. By 1993, this situation had reversed. In some countries, the chemicals industry has grown to become a significant economic sector; in Taiwan, for example, the chemical industry accounted for 30 percent of manufacturing in 1996 versus 10 percent in the United States and Western Europe.

Source: Buccini 2004.

Trade Measures

Countries vary with respect to their vulnerability to climate change and their willingness to pay to avoid any future damage. Because of the difficulties associated with attaining the international cooperation that will be needed to enact effective policies for addressing climate change, there is a widespread concern that countries might start using unilateral measures to address differential attitudes, perceptions, and policy standards. Such unilateral measures may take the shape of trade measures, such as tariffs or quotas against countries that refuse to participate in global efforts on climate change. In the most extreme case, an individual country may unilaterally define standards and then apply sanctions to enforce compliance with those standards.

The theoretical analysis (Baumol and Oates 1988; Copeland 1996; Ludema and Wooton 1994; Mani 1996; Markusen 1975) points to a role for trade restrictions in a second-best setting (environmental taxes being the first-best measures).[5] They suggest use of trade measures either as tools to maximize the welfare of the importing country, if it is directly affected by pollution from the exporting country (as in a transboundary pollution), or as "weapons" to persuade the exporting country to introduce some standard measures of pollution control. It is difficult in such a general framework to get much sense of the appropriate magnitude of such tariffs and their potential effects. Further, much of the focus in the literature is on local pollution in a two-country setting where *transboundary* pollution in one affects the other, as opposed to *global* pollution (like GHG emissions), which affects the entire world. As suggested by the theoretical analysis, no mitigating measures for climate change are in use anywhere in the world.

Ironically, the first legislative proposal on the use of trade policy to address differential environmental standards was introduced in the U.S. Senate.[6] The proposed legislation— called the International Pollution Deterrence Act of 1991— suggested that a countervailing levy or an environmental tariff be imposed against foreign nations whose exports benefited from the cost advantages stemming from less strict environmental standards than those in the United States. The amount of the tariff was proposed to equal the per unit difference between the environmental compliance costs of the United States and its trading partners. While the legislation never made it past the Senate, it once again brings to light the issue of effectiveness of trade measures to address global environmental concerns.[7]

The use of trade measures to enforce compliance with the Montreal Protocol and other such international agreements should not be confused with the proposed Kyoto tariff. The Kyoto tariff targets the United States and other nations that shun the Kyoto agreement or any such future agreements; and it is somewhat similar to the 1991 U.S. proposal. The main purpose of such a tariff would be to protect EU industries from international competition arising from the implementation of the Kyoto Protocol without necessarily addressing the climate change issue in the exporting country.[8]

In spite of the theoretical underpinnings of using trade measures as plausible second-best measures, two important questions warrant greater attention in the climate change debate. First, are trade barriers an appropriate way to address global environmental concerns? Second, if imposed, how are these measures going to affect the patterns of world trade and hence emissions?

Are Trade Measures Compatible with the WTO?[9]

Under the GATT, WTO members can adopt measures to protect the environment and human health and life as long as such measures comply with GATT rules or fall under one of the exceptions to these rules.[10] The most relevant GATT rules for climate policies include the following:

- The *Most-Favored-Nation Obligation* (Article I) requires member states to accord the same treatment to like products produced by other member states—that is, not to discriminate among like products of different member states.
- The *Tariff Obligations* (Article II) require member states to fix tariff levels, and prohibit tariffs above such levels.
- The *National Treatment Obligation* (Article III) prohibits member states from applying "internal taxes and other internal charges, and laws, regulations and requirements affecting the internal sale, offering for sale, purchase, transportation, distribution or use of products . . . to imported or domestic products so as to afford protection to domestic production."
- *Internal Tax* (Article III(2)) prohibits a country from imposing "internal taxes or other internal charges to imported or domestic products in a manner contrary" to the National Treatment Obligation principle in Article III. The section suggests that adjustable *product taxes* (i.e., domestic sales, value added, and excise taxes) can be applied to imports, but not *producer taxes* (i.e., payroll or income taxes, social security charges, or taxes on projects or interests), as long as they are not discriminatory.[11]
- The *Prohibition on Quantitative Restrictions* (Article XI) requires member states to refrain from imposing quotas, including bans, on imports of products from other member states, except in specified circumstances.

In some circumstances, if a country adopts an import ban on products from countries that do not have carbon restrictions or impose punitive import tariffs on such products, such measures could violate WTO rules unless they fall under one of the exceptions.[12]

Article XX provides for possible exceptions to the preceding requirements. Two of these exceptions are of particular relevance to climate change policies. Articles XX(b) and (g) allow WTO members to justify GATT-inconsistent measures if these are either necessary to protect human, animal, or plant life or health, or if the measures relate to the conservation of exhaustible natural resources, respectively. Moreover, the chapeau of Article XX requires that these measures not arbitrarily or unjustifiably discriminate between countries where the same conditions prevail, nor constitute a disguised barrier to trade.

A particularly thorny issue in assessing the compatibility of trade measures with climate change policy may arise with the application of measures based on processes and production methods (PPM). These PPM-based measures may be targeted at the way products are produced, as opposed to the inherent qualities of the product itself. Since most climate change measures do not directly target any particular products, but rather focus on the method by which greenhouse gases may be implicated related to production, PPM issues are critical for the compatibility analysis.

In the Shrimp-Turtle dispute,[13] the WTO Dispute Settlement Panel and the Appellate Body may have opened the doors to the permissibility of trade measures based on PPMs. Previous cases had not been friendly toward the concept of PPMs. However, in that case, India, Malaysia, Pakistan, and Thailand challenged a ban imposed by the United States on the importation of certain shrimp and shrimp products from these countries. The measure at issue involved how shrimp were caught; that is, whether shrimp trawlers used "turtle excluder devices," which allowed shrimp to pass to the back of the net while directing endangered sea turtles and other unintentionally caught large objects out of the net.

The WTO panel and the Appellate Body focused on the manner in which the United States applied its measure, and found that it met the requirements of Article XX exceptions, including the requirements of the Article XX chapeau, which prohibits measures from being applied in an "arbitrary or unjustifiable" manner or used as a "disguised restriction on international trade."[14] They also noted the appropriateness of certain measures in certain circumstances to protect the environment, pointing out that sea turtles were protected under the widely ratified Convention on International Trade in Endangered Species, to which all of the parties to the WTO dispute were also parties. Therefore, even if a climate change policy (including those PPM-based measures) might not be fully GATT-consistent, depending on the circumstances, it may be justified if it meets the requirements of Article XX exceptions, and is not arbitrarily or unjustifiably applied or constitutes a disguised barrier to international trade.

What Is the Impact of Such Measures?

While the distortionary impacts of these environmental tariffs are often hard to predict, it is useful to get some sense of the likely directions and magnitude of some of these effects. We undertook a trade simulation exercise using a partial equilibrium approach to understand the potential impact of an EU "Kyoto tariff" or carbon tax on U.S. exports. The advantage of using a partial equilibrium approach here is that analysis done at the detailed tariff level enables one to make projections for a well-defined set of products (see appendix 5 for a brief description of the model).[15]

We calculated the trade creation effects that would result from the EU-imposed carbon tax on U.S. exports of the most energy-intensive products (pulp and paper, industrial chemicals, nonmetallic mineral products, iron and steel, and nonferrous metals).[16] A range of Kyoto tariffs (10, 20, and 30 percent) was assumed, to reflect to some extent the market price of a ton of carbon in the EU Emissions

TABLE 2.6
Impact of an EU "Kyoto Tariff" on U.S. Exports

	US$ 000s	Loss in Total U.S. Exports (%)	Loss in U.S. Energy Intensive Exports (%)
Total EU imports from the United States (2005)	207,713,157		
EU imports of energy-intensive products from the United States	46,000,809		
10 percent Kyoto tariffs		2.3	10.2
20 percent tariffs		4.5	20.4
30 percent tariffs		6.8	30.5

Trading Scheme (ETS). To calculate the trade creation effects, we derived (from the World Bank trade database) the data on imports and corresponding import elasticity of demand for products at the country level.

The results discussed in table 2.6 suggest that the United States would lose up to 7 percent of its exports to the EU if such tariffs were implemented. But the energy-intensive industries, such as the steel and cement industries (which will be subject to this tariff), would be most severely affected and could suffer up to a 30 percent loss. Even this is an underestimation, as it does not take into account the trade diversion effects that could result from the EU shifting to other trading partners whose tariffs now are much lower than the tariff on the United States. The simulation exercise thus suggests that if the EU goes ahead with the proposal to introduce border taxes to compensate for the climate change policies, it could significantly affect the U.S. trade balance. We did not attempt a similar exercise for developing countries (e.g., China), since the current EU debate is still an industrial-country issue mainly targeted at the United States.

As discussed in the previous section, there are still issues with regard to the WTO compatibility of these punitive measures.[17] The recent WTO panel ruling on the "Shrimp-Turtle" case seems to have at least started a debate for considering PPM measures as long as they are not imposed in a discriminatory fashion. Even considering the practical costs and implementation hurdles, the environmental benefits and impacts of a Kyoto tariff should not be underestimated.

WTO and Kyoto Protocol: Exploring Synergies for Advancing Both Trade and Climate Agendas

As more and more countries move toward adopting climate-friendly policies, the economic and trade ramifications are likely to bring increasing attention to the relationship between the trade and climate regimes.

In its Preamble to the Marrakesh Agreement, which established the WTO in 1995, the WTO recognizes the importance of seeking to "protect and preserve the environment." The Kyoto Protocol states that parties should "strive to implement policies and measures in such a way as to minimize adverse effect on international trade." The UNFCCC features similar language in several places (Frankel 2004), and the Doha *Communiqué* specifically states that "the aims of upholding and safeguarding an open and nondiscriminatory multilateral trading system, and acting for the protection of the environment and promotion of sustainable development can and must be mutually supportive."

There is thus a general recognition by both regimes to respect the other's mandate. Further, the Doha round contains specific provisions that could promote Kyoto objectives. For example, a multilateral liberalization of environmental goods and services, such as air quality improvement and climate policy (e.g., windmill turbines), would serve both kinds of goals—economic and environmental. While not very explicit in the current WTO discussions, a ban on subsidies to fossil fuels (akin to the discussion on fisheries subsidies) would achieve both the environmental goal of reducing carbon emissions and the goal of removing an economic distortion.[18]

In light of these issues, the potential conflict between climate change mitigation under the Kyoto Protocol and the system of trade rules under the WTO has drawn much attention recently among academics and policy makers, and it has spawned much discussion on how best to avoid such conflicts.[19] In the future, both the climate change regime and trade investment regime will ideally evolve to accommodate new economic and political circumstances. It is therefore important to continue to monitor and analyze the relationships between the two regimes. There is much to be gained by working together to achieve common goals of climate policy and development, especially given the increasing number of developing countries that will also come into play in the coming years.

It would be useful at the outset to focus on a few areas where synergies could be exploited in the immediate short run. The energy efficiency and renewable energy technologies needed to meet future energy demand and reduce GHG emissions below current levels are largely available. As discussed in the coming chapters, WTO parties can do their part by seriously considering liberalizing trade in climate-friendly and energy-efficient goods as a part of the ongoing Doha negotiations to support Kyoto. Within the Kyoto Protocol, the most important priority regarding the linkage to trade would be to facilitate a uniform approach to taxation of energy and greenhouse gas emissions. Such an approach would eliminate conditions of competitiveness and leakage now resulting from uneven treatment across countries.

Nordhaus (2007) argues that if carbon prices are equalized across participating countries, there will be no need for tariffs or border tax adjustments among participants. While much work on the details would be required, he suggests that this is a familiar terrain because countries have been dealing with problems of tariffs, subsidies, and differential tax treatment for many years (through the WTO). The issues, according to Nordhaus, are elementary compared to the complexities of a quantity-based regime as in the Kyoto Protocol. The Protocol also specifically mentions "progressive reduction or phasing out of market imperfections and subsidies in all greenhouse gas emitting sectors" as one of the measures that parties could adopt to help achieve their emission targets. The issues also are consistent with multilateral trading arrangements.[20] Equalizing carbon prices will also help avoid the perception and reality that climate measures might be used as an excuse for protectionist discrimination.

Key Findings from Chapter 2

- **A variety of regulatory and fiscal measures to combat climate change are already in place in a number of OECD countries.**
- **Analysis suggests that energy efficiency standards adversely affect the competitiveness of industries more than carbon taxes do.**
- **Evidence shows that some leakage of carbon-intensive production to developing countries is already happening.**
- **Simulation exercises suggest that an EU-imposed "Kyoto tariff" will adversely affect U.S. exports, especially industries such as steel and cement.**
- **The WTO principles and Kyoto Protocol contain mutually compatible areas that could be further developed for the benefit of trade and climate change.**

Notes

1 Of the three major fossil fuels, coal produces the most carbon per unit of energy, followed by oil and then natural gas.

2 There are several estimates relating to the impacts of a CO_2 tax on GHG emissions. According to the Nordic Council of Ministers (2002), CO_2 emissions in Denmark decreased 6 percent during the period 1988–97, when the economy grew by 20 percent. They also decreased 5 percent just in 1996–97, when the tax rate was raised. A similar study in Norway, on the other hand, suggested that a carbon tax resulted in only a 2 percent reduction in emissions. A study of the climate change levy in the United Kingdom (Cambridge Econometrics 2005) revealed that total CO_2 emissions were reduced by 3.1 mtC (million tons carbon)—2 percent—in 2002 and by 3.6 mtC in 2003 compared with the reference case.

3 A pollution haven may arise if environmental stringency differs between countries, when capital is mobile, and when trade rules allow firms to relocate and still sell their products to the same customers. A general consensus from the literature is that any tendency toward formation of a pollution haven is self-limiting, because economic growth brings countervailing pressure to bear on polluters through increased regulation (Mani and Wheeler 1998).

4 It is now well documented that the energy shocks of the 1970s and 1980s were responsible for some of the relocation of energy-intensive industries from developed to developing countries, which at that time still had huge subsidies in place.

5 There are two major reasons why an import tariff will not be as efficient as a Pigouvian environmental tax. First, unlike a Pigouvian tax, a tariff does not directly affect the cost of the polluting product: it works indirectly by influencing demand. Second, a tariff by an importing country could possibly reflect only those detrimental effects that fall within its borders and hence does not take into account the overall externality generated by the production process.

6 One argument is that around that time (1991), the United States was quite active in the international environmental field, often more progressive than the EU. There is much discussion of this in the recent book of Philippe Sands (2005), titled *Lawless World: America and the Making and Breaking of Global Rules.*

7 Mani (1996) showed that an environmental tariff introduced in this fashion will have no significant impact on the patterns of world trade and pollution.

8 Trade controls have been employed to ensure compliance in a number of multilateral environmental agreements—such as hazardous waste, fisheries, endangered species, and ozone depletion—over many decades. By contrast, a trade sanction is a specific action to coerce governmental behavior. The only two international organizations that impose trade sanctions against noncompliance are the UN Security Council and the WTO.

9 Determination of WTO compatibility, of course, is in the jurisdiction of the WTO. This section is intended only to highlight some of the relevant GATT rules that may be triggered by a country's trade measures to address climate change. For detailed analysis of WTO compatibility and discussion on this subject, see Pauwelyn (2007), Petsonk (1999), Werksman (1999), and Zhang and Assuncao (2004).

10 This right to adopt environment-related measures has been affirmed by several panels and the Appellate Body. See, for example, *US – Gasoline* (D52), "WTO Members have a large measure of autonomy to determine their own policies on the environment (including its relationship with trade), their environmental objectives and the environmental legislation they enact and implement. . .[and] that autonomy is circumscribed only by the need to respect the requirements of the *General Agreement* and the other covered agreements"; http://www.wto.org/english/tratop_e/envir_e/gas1_e.htm. Also see *EC – Asbestos* (T.4.1.1): WTO Members have the "right to determine the level of protection of health that [it] consider[s] appropriate in a given situation"; http://www.wto.org/english/tratop_e/dispu_e/repertory_e/t4_e.htm.

11 The question is whether a carbon tax or any other taxes would be considered product taxes or producer taxes and whether the obligation to hold emission credits or allowances would be considered "internal taxes or other internal charges" under Article III(2).

12 A new proposal to establish a mandatory U.S. cap-and-trade system is gaining support in the context of legislation being considered in the U.S. Congress. The system would require, in the future, importers to purchase emission allowances to offset imports into the United States from China, India, Brazil, and other countries. The proposal is gaining a lot of political support in the United States, especially among the labor and environmental groups.

13 *United States—Import Prohibition of Certain Shrimp and Shrimp Products*, WT/DS58/AB/R.

14 *United States—Import Prohibition of Certain Shrimp and Shrimp Products—Recourse to Article 21.5 of the DSU by Malaysia*, WT/DS58/AB/RW, paras. 153–154. The Panel and the Appellate Body found that the U.S. measure, as modified, "no longer constitute[d] a means of unjustifiable or arbitrary discrimination" because (i) the U.S. had made serious, good faith efforts to negotiate an international agreement and (ii) the revised guidelines required that other Members' programs simply be "comparable in effectiveness" to the U.S. program, as opposed to being "essentially the same." The Appellate Body conditioned this finding,

however, by stating that the U.S. measure was justified under Article XX "as long as the[se] conditions . . . in particular the ongoing serious good faith efforts to reach a multilateral agreement, remain satisfied." Ibid. at para. 153.

15 Partial equilibrium trade simulation models are widely used in the literature to estimate the effects of change in tariffs and nontariff barriers (see Laird and Yeats 1990).

16 *Trade creation* refers to change in overall demand for imports whose price has changed relative to domestic substitutes.

17 For a detailed exposition of this issue, see Bhagwati and Mavroidis (2007).

18 Presently, the WTO contains no special provisions relating specifically to these subsidies. This means that these subsidies are disciplined only by the general subsidies rules found in the current WTO Subsidies Agreement (SCM Agreement).

19 There is a rich literature on this subject, and hence it is not discussed here. For the most comprehensive assessments see Brewer (2003), Charnovitz (2003), Cosbey (2003), and Frankel (2004).

20 Article 2.1 of the Kyoto Protocol, cited in Brewer (2003).

CHAPTER 3

Beyond Kyoto: Striving for a Sustainable Energy Future in Developing Countries

WHILE OECD COUNTRIES WILL REMAIN the largest per capita emitters of greenhouse gases, the growth of carbon emissions in the next decades will come primarily from developing countries, which are following the same carbon-intensive development path that their rich counterparts did. Among the developing countries, it is expected that the main growth in carbon emissions will emanate from China and India because of their size and growth. The International Energy Agency (IEA) projects that between 2020 and 2030, developing country emissions of carbon from energy use will exceed those of developed countries in aggregate, but they will still lag far behind on a per capita basis.

Given that it is aggregate emissions that count toward global warming, and these have historically come from OECD countries, the UNFCCC has recognized the concept of "common but differentiated responsibilities." This concept has been built into the Kyoto Protocol and the trading of carbon emission reduction credits under the Clean Development Mechanism (CDM). However, this is likely to change in the post-Kyoto scenario, because developing countries like China and India might increasingly be called upon to meet global emission reduction targets.

This chapter is organized as follows. The first section provides an overview of the global trends of increasing greenhouse gas emissions and sets the stage for developing countries to consider policy options that can reconcile the trade and climate agenda. Using trade data on selected high-GHG-emitting countries, the next section describes the role and evolution of low-carbon technology in the context of climate change mitigation. The chapter then describes the existing tariff and nontariff barriers to the use of climate-friendly technologies in these countries and assesses the trade differential from the changes in tariff and nontariff barriers across two scenarios. The final section summarizes the main findings.

Global Emissions Scenarios through 2030

As discussed earlier, global emissions of greenhouse gases have continued to rise during the last two decades. According to the International Energy Agency, world emissions of CO_2 from fossil fuel combustion increased from 20.8 billion tons (Gt) in 1990 to 26.6 Gt in 2004, an increase of 28 percent.[1]

While the largest share of historical and current global emissions of greenhouse gases has originated in developed countries, developing countries will soon account for a greater share of world CO_2 emissions from fossil fuel combustion than developed countries. The IEA's projections suggest that based on energy use, non–Annex I countries will overtake the Annex I countries as the leading contributor to global emissions in the 2020s. Non–Annex I countries' share of global emissions will soar from 38 percent in 2002 to 52 percent in 2030, while Annex I countries' share will decline from 60 percent to 47 percent (figure 3.1).[2]

In other words, more than 70 percent of the global emissions increase from 2020 to 2030 will come from non–Annex I countries (table 3.1). China alone will contribute about a quarter of the increase in CO_2 emissions, or 3.8 Gt, reaching 7.1 Gt in 2030. Its emissions will overtake those of the United States by 2010

FIGURE 3.1
CO_2 Emissions from Energy Use, 2002-30

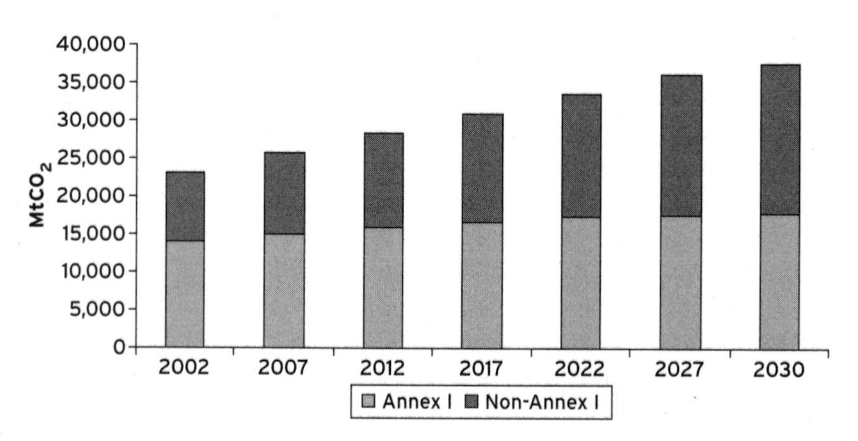

Source: IEA database (2006).

(IEA 2006). Strong economic growth and heavy dependence on coal in industry and power generation contribute to this trend. India is also contributing to the increase in global emissions; it is projected to add about 8 percent of the total increase in emissions, or 1.2 Gt, by 2030.

It is clear that a future international framework for climate change needs the participation of developing countries, particularly major emitters, in order to have a real impact on global emissions trends. Development is likely to lead to increasing demand for energy, and without adequate climate policies in developing countries, producers and consumers in those countries will not modify their behavior to reduce climate change risks (Stern 2006).

TABLE 3.1
Potential Contribution to CO_2 Increase, 2002-30

	Percentage of the Increase
Annex I	26
Non-Annex I	76
(of which)	
China	26
India	8
Indonesia	3
Brazil	3
Mexico	2

Source: WRI Climate Analysis Indicator Database (CAIT).

The Kyoto Protocol is an important first step toward international cooperation to deal with the challenge of climate change. However, the Protocol is weakened because not all countries with obligations to reduce their emissions have ratified the agreement, and because, at least at this stage, it does not impose commitments to reduce emissions on the major developing-country emitters. In response to their historical responsibility and financial and technological capabilities, only developed countries (Annex I countries) were required to adopt fixed emission targets under the Kyoto Protocol.

Most Annex I countries have implemented various polices and measures to achieve their targets and showed some progress in enacting measures to mitigate climate change. However, in a number of cases, economic considerations have far outweighed considerations for the global climate. Many of the incentives, especially for energy-intensive industries to reduce their emissions, have been nullified through special tax concessions, rebates, exemptions, and other such measures. Nonetheless, whatever emission reductions have been achieved in developed countries is likely to be largely offset by growth in developing countries. It is thus critical that all countries collectively identify cost-effective policies and measures they can enact to contribute to substantial and long-term reductions in greenhouse gas emissions. This chapter explores trade policy as one such option in the context of clean energy technology transfer to developing economies.

Clean Energy for the Future

With some emerging economies growing at 8 to 10 percent, their demand for energy is expected to increase three to five times by 2050 (World Bank 2006a). As carbon-intensive energy infrastructure and cities are being rapidly built and expanded, there is little emphasis on cleaner and more efficient technologies. Although OECD countries will remain the largest per capita emitters of greenhouse gases, the growth of carbon emissions in the next decades will come primarily from developing countries. Bringing down the potential growth in GHG emissions will require that steps be taken on two fronts:

- Improving energy efficiency on the demand side; and
- Investing in technologies on the supply side (e.g., electricity generation) to increase efficiency and reduce carbon emissions.

Technology Transfer in the Context of the Kyoto Protocol

In response to developing country needs, the UNFCCC identifies provisions related to technology transfer across five themes: technology needs and needs assessments, technology information, enabling environments, capacity building, and

BOX 3.1

Approaches to Employing Technology Investments in Developing Countries

International technology transfer through trade occurs when a country imports higher-quality intermediary goods (than it can produce itself)—such as steam turbines and boilers—to use in its coal combustion processes. The study by Hakura and Jaumotte (1999; cited in OECD 2002), using data from 87 countries, concludes that trade serves as a channel for international technology transfer to developing countries. However, it appears that intraindustry trade plays a more important role in technology transfer than interindustry trade. Intraindustry trade is more pervasive among developed countries, and interindustry trade is more prominent in trade between developed and developing countries. Hence, an immediate implication of their findings is that developing countries will enjoy relatively less technology transfer from trade than developed countries. Because of this finding, we are led to consider other approaches employed by developing countries to acquire technologies. These channels, which are closely interrelated and support each other, include the following:

INVESTMENT. A firm can set up a foreign establishment to exploit the technology itself. Foreign direct investment (FDI) is the most important means of transferring technology to developing countries. Technology transfer through FDI generates benefits that are unavailable when using other modes of transfer. For example, an investment comprises not only the technology, but also the entire "package," such as management experience, entrepreneurial abilities that can be transferred by training programs, and learning by doing. Further, many technologies and other know-how used by affiliates of multinational enterprises (MNEs) are not always available in the market, but only through the MNE itself. And some technologies, even if available in the market, may be more valuable or less costly when applied by the firm that developed them rather than by an outsider. Similar to FDI, but not conferring the same level of control to the parent investor, is a range of cooperative arrangements, including joint ventures, subcontracting, and franchising.

mechanisms for transfer.[3] The enabling environment component is particularly useful for this paper, as the focus here is on government actions—such as fair trade policies; removal of technical, legal, and administrative barriers to technology transfer; sound economic policies; regulatory frameworks; and transparency—that create an environment conducive to private and public sector technology transfer.

Trade Issues Related to Clean Energy Technologies

As discussed in the chapter 1, a number of low-carbon technologies already exist to combat climate change. Thus, international technology transfer can be a significant and cost-effective component of climate mitigation efforts (box 3.1). In this chapter, trade issues related to some key clean energy technologies are explored. Given the broad range of clean energy technologies, a detailed analysis of all the

INTERNATIONAL JOINT VENTURES. IJVs are also a common business arrangement for international technology transfer because firms in different countries exploit opportunities for mutually beneficial and complementary international interfirm transfers. An IJV involving a local firm is sometimes required by the host government as a condition of doing business in the host country. IJVs are also a common business arrangement for the purpose of international technology transfer as firms in different countries exploit opportunities for mutually beneficial and complementary international interfirm transfers. An IJV involving a local firm is sometimes required by the host government as a condition of doing business in the host country.

LICENSES. A firm may license its technology to an agent abroad who will use it to upgrade its own production. Successful penetration of foreign markets can seldom be based on exports alone. Various tariff and nontariff barriers, government policies, or the general investment climate can make exporting a costly option. Also, for certain industry sectors, notably in services, trade can be a complicated means to exploit a firm's superior technology or management capabilities overseas. In those cases, a firm may choose to license its technology to a local firm.

TEMPORARY RELOCATION OF EMPLOYEES. Technology is often transferred internationally by employees of multinational firms or through the migration of individual experts.

INTERNATIONAL DEVELOPMENT AID. Several countries have put in place initiatives under their national development programs to facilitate the transfer of clean coal technologies to developing countries. For example, Japan has an initiative that aims at promoting and accelerating the introduction and dissemination of technologies for energy savings. The United States has developed a clean technology initiative that focuses on clean coal technology.

Source: OECD 2002.

technologies is beyond the scope of this study. We conducted case studies relating to four technology groups—high-efficiency and clean coal technologies, efficient lighting, solar photovoltaics, and wind power—to examine the issues involved in promoting increased international trade in clean energy technologies.

These technologies were selected for three main reasons. First, they constitute low-carbon-growth strategies in many developing and developed countries. Second, the choice of technologies identified for the current study is consistent with the World Bank's clean energy investment framework. Finally, the choice of these technologies is also reflected in current WTO negotiations on environmental goods and services. In those negotiations, nine members (Canada, the European Communities, Japan, Rep. of Korea, New Zealand, Qatar, Switzerland, Chinese Taipei, and the United States) have tabled submissions containing their initial lists of environmental goods, including a wide range of clean energy technologies for reducing trade and nontrade barriers (WTO 2005).

Technology Codification

Global trade is typically tracked based on a unique Harmonized Commodity Description and Coding System (harmonized system, or HS) for each commodity. The harmonized system contains over 5,000 product codes. Under the system, each product traded is assigned a six-digit code. To track the volume of trade in clean energy technologies and the corresponding tariffs levied across countries, this study used the six-digit HS code developed and updated by the World Customs Organization. Typically, each component of a particular technology should be associated with a different HS code. In addition to the six-digit system, regions and countries may have their own systems to define the products more specifically with eight- or even 10-digit codes (box 3.2).

At a six-digit HS code level, clean energy technologies and components are often found lumped together with other technologies that may not necessarily be classified as environmentally sustainable or clean technologies. Consequently, data for clean energy technologies relating to international trade may be overestimated or underestimated, resulting in a possible limitation of this study. An example is that solar photovoltaic panels are categorized as "Other" under the subclassification for light-emitting diodes (LEDs) under the HS codes. Such a categorization suggests that reducing the customs tariff on solar panels might also result in tariff reduction for unrelated LEDs (Steenblik 2006; Vernstrom 2007). Similarly, technologies relevant for clean coal electricity generation and for cleaner industrial use are not clearly classified under a separate HS category, which makes them difficult to track.

The imprecise definition of clean technologies across HS codes also raises another issue for countries that are considering lowering trade barriers for clean energy equipment and related components. In cases in which the codes

BOX 3.2

Regional and Country-Specific HS Nomenclature

The ASEAN Harmonized Tariff Nomenclature (AHTN) adheres to the HS code, but includes two additional digits for more precise definition. ASEAN permits member countries to add digits to the existing AHTN classifications for domestic purposes. The United States has adopted a 10-digit classification system based on the HS codes to allow for more detailed product specificity within each eight-digit classification. Similarly, India has defined renewable energy goods more precisely within the HS system of codes (well beyond the standard six digits). For example, the customs duty for wind energy-related equipment and components has been described in detail and codified as demonstrated below.

Customs Duty for Wind Energy Equipment and Components in India

Item	Customs Duty Rates: 2002-03			
	Basic (%)	Surcharge	Additional	Special Additional (%)
Wind-operated electricity generators up to 30 kW and battery chargers up to 30 kW	5	Nil	Nil	4
Parts for manufacture of wind-operated electricity generators, namely, (a) Special bearings (b) Gearbox (c) Yaw components (d) Sensors (e) Brake hydraulics (f) Flexible coupling (g) Brake calipers (h) Wind turbine controllers (i) Parts of goods specified at (a) to (h) above	5	Nil	Nil	4
Blades for the manufacture of rotor of wind-operated electricity generators	5	Nil	Nil	4
Parts for the manufacture or the maintenance of blades for rotor of wind-operated electricity generators	5	Nil	Nil	4
Raw materials for the manufacture of blades for rotor of wind-operated electric generators	5	Nil	Nil	4

Source: India's Ministry of Non-conventional Energy Sources 2004.

are not detailed enough, the scope of the tariff reduction becomes much broader than necessary. In countries where a large proportion of the tax revenue comes from international trade, the challenge faced by the government becomes more complex as a government's ability to consider special breaks for clean energy is constrained, especially if clean technologies are lumped together with other technologies (see chapter 4).

Liberalization of Trade in Clean Energy Technologies

Within the constraints presented above, this section discusses the impact of reduced tariff and nontariff barriers on trade volumes, which can be analyzed for four specific clean energy technologies identified for the study. As described in chapter 2, information regarding trade flows is available through the WITS (UN Comtrade's World Integrated Trade Solution) database. Trade simulations are carried across two scenarios for four specific technologies based on the partial equilibrium model presented in detail in appendix 5:

1. Clean coal technologies (HS codes 840510, 840619, 841181, 841182, 841199)
2. Wind energy (HS codes 848340, 848360, 850230)
3. Solar photovoltaic systems (HS codes 850720, 853710, 854140)
4. Energy-efficient lighting (HS code 853931)

Scope of the Study

The HS code data associated with each technology includes requisite key components associated with each clean technology. These components are discussed in detail in this chapter's corresponding section for each technology. It must be highlighted, however, that the data tracked under codes associated with clean coal may involve other dual-use components that cannot necessarily be justified under clean technologies or components.

Data

Trade data are analyzed for the top 18 developing countries based on their GHG emissions.[4] The most recent complete trade information available for all countries is for 2004. Levels of tariffs are available through the WITS database. The data on nontariff barriers (NTBs) are derived from the World Bank's own trade database (Kee, Nicita, and Olarreaga 2005). The NTBs are calculated by transforming all the information on NTBs into a price equivalent. The ad valorem equivalent (AVE) of the core NTBs thus calculated includes price and quantity control measures, technical regulations, as well as monopolistic measures, such as a single channel for imports.

To study the effects of tariffs, one needs import demand elasticity data at the tariff line level that are consistent with GDP maximization. Import demand elasticity data used here are derived from the World Bank's Global Monitoring Report database and measures the percentage change in import volume due to a 1 percent increase in import price. The database contains import elasticities for over 4,625 goods (at the six-digit level of the Harmonized System) in 117 countries using a methodology that is consistent with trade theory (i.e., imports are a function of prices and factor endowments). Data sources are identical for all countries and goods.

Two different scenarios used here analyze the liberalization of clean energy technologies. The first scenario assesses the change in trade volume of clean energy technologies when tariffs are completely eliminated across all 18 high-GHG-emitting developing countries for the four aforementioned technologies. The second scenario assesses the change in trade volume of these technologies when both tariffs and NTBs (calculated as ad valorem equivalents) are completely removed across the same sample of countries. As this analysis is based on a limited set of HS codes, it would need to be validated based on a more thorough analysis using a wider set of technologies and larger set of countries.

Conclusions

By eliminating tariff and nontariff barriers in 18 high-GHG-emitting developing countries, trade liberalization results in huge gains in trade volumes, as illustrated in table 3.2. It is worth noting that the changes in trade volumes, which range from 3.6 percent to 63.6 percent across the four technologies identified for the study, result from the varied level of tariffs on the technologies; the nontariff barriers, namely quotas and technical regulations; other investment barriers related to intellectual property rights; and the import elasticity of demand for these products. The assessment is based on first-round approximations rather than full general equilibrium effects that would be important in the context of global trade. Accounting for these second-round impacts would require a full, global general equilibrium model, which is far beyond the scope of this study.

Trade and investment barriers related to each technology are discussed in detail later in the chapter.

TABLE 3.2
Change in Trade Volumes in High-GHG-Emitting Developing Countries from Liberalizing Clean Energy Technologies

Technology Option	Liberalization Scenario 1 (%) Elimination Tariff (only)	Liberalization Scenario 2 (%) Elimination Tariff and Nontariff Barriers
Clean coal technology	3.6	4.6
Wind power generation	12.6	22.6
Solar power generation	6.4	13.5
Efficient lighting technology	15.4	63.6
All 4 Technologies	7.2	13.5

Clean Coal Technology

The major developing countries are following the same energy-intensive growth paths involving the use of coal as their richer counterparts have done. In fact, current global coal demand already lies above earlier forecasts for 2030, with no signs that the growth trend will reverse. The rate of growth is significant in almost all regions and countries, except in North America and Europe. China and India have added significant coal-fired capacity to meet projected demand: 27.5 gigawatts (GW) per year in China (2000–05) and 1.6 GW per year in India over the same period. According to China's National Development and Reform Commission, over 50 GW of new coal-fired capacity should come on-line in 2006. India is likely to fall short of its 3.5 GW target for 2006.

The use of clean coal technologies is critical for non–Annex I countries, specifically China and India, where the load of carbon emissions results from thermal power generation and industrial expansion (figure 3.2). There is tremendous scope for upgrading existing coal combustion systems to foster cleaner production mechanisms (box 3.3). This study focuses on electricity generation technology with particular emphasis on coal combustion.

Coal Combustion Technology (IGCC) with Increasing Climate Benefits

During an initial scoping exercise, the study focused on two mechanisms for coal combustion: supercritical and ultra-supercritical boilers and turbines in pulverized coal thermal power generation, and integrated coal gasification combined cycle (IGCC). However, it was not possible to identify a six-digit HS code that

FIGURE 3.2
Energy Production in China and India, 2004

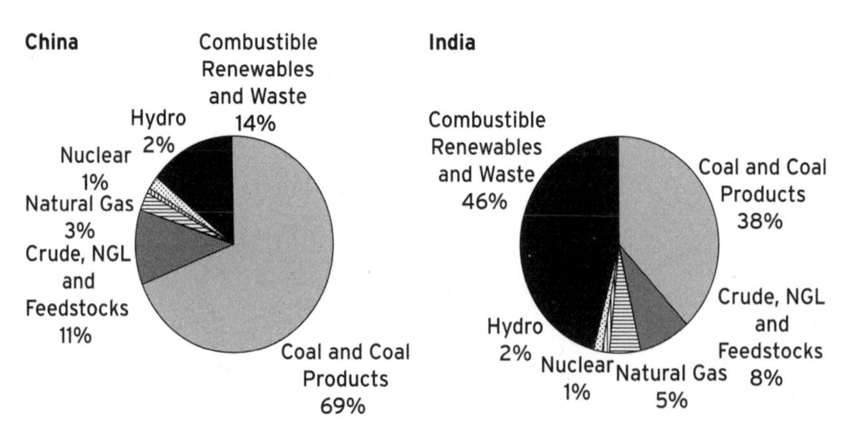

Source: IEA (2006).

BOX 3.3

Clean Coal Technologies

Clean coal technology refers to various technologies that aim to improve energy efficiency and reduce environmental impacts, including technologies of coal extraction, coal preparation, and coal utilization. Clean coal technology can be categorized differently from different perspectives. The International Energy Agency (IEA) divides clean coal technology into the following categories:

COAL EXTRACTION AND PREPARATION TECHNOLOGY. The technology includes reliable and high-efficiency modern coal extraction technology and modern coal preparation technology, which could greatly decrease ash and remove impurities such as sulfur. This category includes technology of coal homogenization, coal preparation, and coal washing, etc.

ELECTRICITY GENERATING TECHNOLOGY. This group of technologies includes high-efficiency combustion technology such as supercritical and ultra-supercritical pulverized coal combustion technology, fluidized-bed combustion technology, pressurized fluidized-bed combustion technology, and integrated coal gasification combined cycle technology, etc.

OTHER USES. Clean coal technology can apply to other industrial sectors such as steel, cement, process/space heating, and many kinds of chemical processes. It can also be applied to households such as heating and cooking, for example, briquette and coal-water mixture technology.

Source: IEA.

could easily serve as a proxy for supercritical and ultra-supercritical boilers (and turbines). As such, the study was streamlined to focus on the IGCC.

The IGCC combines coal gasification and combined-cycle power generation technologies. Coal gasification converts solid coal into a combustible gas composed primarily of carbon monoxide and hydrogen. The gas is then cleaned of sulfur compounds and particulate matter and burned in a gas turbine to generate a first source of electricity. Exhaust gas from the gas turbine is used to produce steam to drive a steam turbine to generate a second source of electricity. The main beneficial features of IGCC are that (a) the gasified coal is purified of sulfur and particulate pollutants before it is burned in the turbine, and (b) the residual heat in the hot exhaust gas is further utilized in a heat recovery steam generator to produce additional electricity and thereby increase the thermal efficiency. The thermal efficiency of IGCC is 42 to 44 percent compared to 35 percent efficiency for existing petroleum coke (PetCoke) plants that do not employ supercritical or ultra-supercritical technology.

The typical size of IGCC power plants is 200 to 500 megawatts. IGCC types may be different for different designs of coal. Modular designs are evolving for future IGCC power plants, with larger sizes integrating multiple units. IGCC plants can burn any high-hydrocarbon fuel, including low- and high-sulfur coal, anthracite, and biomass.

There are presently several commercial-scale IGCC plants in operation in the world. Some of these projects are in the United States and were implemented with the financial support of the U.S. Department of Energy's Clean Coal Technology Program. Two other plants are in Europe: one in the Netherlands and one in Spain. The U.S. plants have General Electric (GE) gas turbines. The European plants use Siemens gas turbines. All IGCC plants in operation are of 250 MW capacity except the unit in Spain, which has approximately 300 MW of capacity.

Market Trends for IGCC

As discussed above, the set of technologies identified for coal combustion also face a problem of imprecise HS codes. Efficient supercritical and ultra-supercritical boilers (and turbines) cannot be easily tracked because there is no suitable HS code differentiation for boilers (and turbines) by temperature and pressure.

Given the limitations in the HS codification system, specific technology components identified below serve as a proxy for IGCC in developing countries, as all the codes available for analysis fall under the dual-use category:

- Producer gas generators
- Steam and vapor turbines over 40 MW
- Steam and vapor turbines not exceeding 40 MW
- Gas turbines not exceeding 5,000 kW
- Gas turbines exceeding 5,000 kW
- Parts of gas turbines

In addition to the five identified IGCC components for coal combustion, also critical are emission control technologies such as particulate removal filters and electrolytic precipitators, flue gas desulfurization (FGD) for reducing sulfur emissions, and NOx control devices. Components such as air separation units and gas cleanup systems are also employed and integral to an IGCC plant. These are all dual-use components with wider applications in chemical and refining industries.

Currently, the main components for coal combustion technology are being produced in the United States, Germany, and Japan and exported to developing countries. The leading producers of gasifiers, steam and gas turbines, and end-of-pipe technologies are GE, Shell, Conoco Philipps, and Siemens for gasifiers, and GE, Siemens, Alstom, and Mitsubishi for gas turbines, to name a few. Table 3.3 provides a list of major exporters and importers for components that can be broadly classified under clean coal technologies. The table shows that China is emerging as one of the major importers of this technology.

TABLE 3.3
Top 10 Trading Countries for IGCC (Clean Coal) Technology Components

	Exporters	Importers
1	United States	United States
2	United Kingdom	Germany
3	Germany	United Kingdom
4	Italy	Iran, Rep. of
5	Switzerland	China
6	Japan	Saudi Arabia
7	France	Italy
8	Mexico	Japan
9	Netherlands	France
10	Hungary	Norway

Source: WITS database.

The trading trends using import-export ratios between high-income OECD and low- and middle-income developing countries between 1995 and 2006 (figure 3.3) suggests that developed countries are still major exporters of clean coal technology (import-export ratio less than 1). Developing countries remain net importers, though the ratio shows a declining trend. The underlying hypothesis for progressively lower imports (or higher exports) of clean coal technologies in developing countries could very well have to do with other investment approaches employed by key participating beneficiaries in energy technology transfers that circumvent trade barriers. This is particularly relevant for China, where FDI and increasing joint ventures are leading to gasification and combined-cycle technology investments in the fertilizer industry (Jin and Liu 1999).

Liberalization of IGCC (Clean Coal) Technologies

This section assesses the existing tariffs and NTBs to IGCC technology in 18 high-GHG-emitting developing countries selected from a list of non–Annex I countries.

FIGURE 3.3
Clean Coal Technology Import-Export Ratio in High-Income versus Low- and Middle-Income Countries

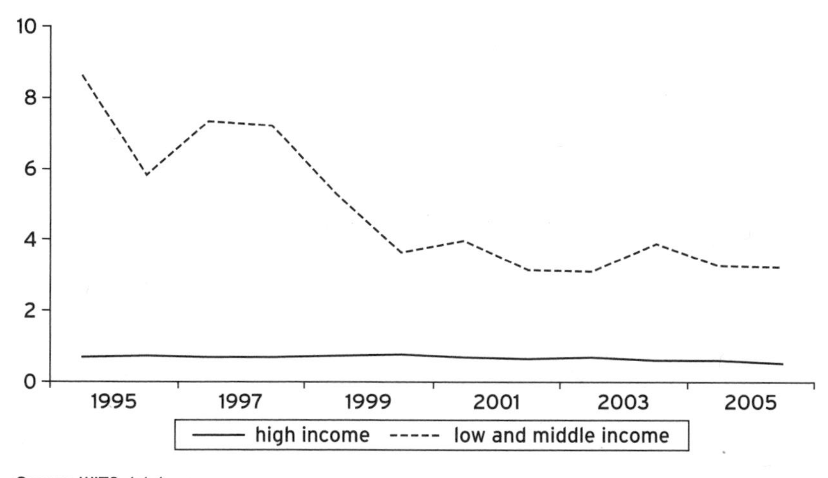

Source: WITS database.

On assessing the tariff levels across 18 countries, the study finds that, with the exception of four countries—Argentina, Indonesia, Kazakhstan, and South Africa—all countries have applied maximum tariffs to at least one of the proxy technologies (table 3.4). The variability in tariffs on specific clean coal technologies is high for some countries and ranges from 12 percent in Mexico to as high as 15 percent in India. The average industrial tariffs are presented here for comparison. Only in some cases do we find that the tariffs on clean coal technology are higher than the average industrial tariffs. One should keep in mind that codification problems prevent examination of the exhaustive list of available IGCC technologies.

As discussed earlier in this chapter, NTBs in the form of quotas and import ceilings applied across these countries are translated into ad valorem equivalents and included in the analysis as an ad valorem equivalent for additional tariffs. Table 3.4 also shows that seven of the 18 countries levy NTBs that range from 160 percent in Nigeria to 25 percent in China. For all other countries, NTBs are almost nonexistent. For comparison, the table also presents the average industrial tariffs and average tariffs in high-income OECD countries.

TABLE 3.4
Applied Average Tariffs and NTBs for IGCC (Clean Coal) Technologies in the 18 High-GHG-Emitting Developing Countries (%)

Countries	Average Tariffs on IGCC Technology	Average Industrial Tariffs	NTBs on IGCC Technology
China	15	10	25
Colombia	15	12	
India	15	29	
Venezuela	15	12	
Brazil	14	14	145
Mexico	12	17	
Bangladesh	6	18	
Chile	6	6	
Zambia	5	12	
Egypt	5	13	149
Nigeria	5	27	160
Philippines	3	6	119
Thailand	1	16	
Argentina	0	12	
Indonesia	0	7	
Kazakhstan	0	3	
Malaysia	0	9	93
South Africa	0	8	125
High-income OECD countries	1	4	

Source: WITS database.

While the impact of tariffs and other cost factors on technology transfers varies across markets and depends largely upon the tariffs applied, the scenario highlighted here illustrates that liberalizing trade can encourage clean coal technology transfer. However, this result does not capture all the other unquantifiable barriers. Trade-related intellectual property rights regimes and investment barriers significantly affect technology diffusion but are not reflected in tariff or nontariff values. Encouraging technology transfer needs other policy measures, such as protecting intellectual property rights and complying with licensing and royalty agreements. Box 3.4 describes a case in China where an impediment to the expansion of clean technology markets exists on account of lax environmental standards and a weak intellectual property rights regime.

Wind Power Technology

Wind power technology is one of the fastest-growing clean energy technologies. According to the Global Wind Energy Council (GWEC), 15,197 MW of capacity was added in 2006, taking the total installed wind energy capacity to 74,223 MW, up from 59,091 MW in 2005. The countries with the highest total installed capacity are Germany (20,621 MW), Spain (11,615 MW), the United States (11,603 MW), India (6,270 MW), and Denmark (3,136 MW). Thirteen countries around the world can now be counted among those with over 1,000 MW of wind capacity, with France and Canada reaching this threshold in 2006 (http://www.gwec.net, 2007).

Market Trends in Wind Power Technology

The wind power market has historically been dominated by dedicated wind-turbine manufacturing companies. More recently, large equipment manufacturers like GE and Siemens have entered the wind power market by acquiring other companies. The top six manufacturers are Vestas (Denmark, merged with NEG Micon in 2004), Gamesa (Spain), Enercon (Germany), GE Energy (United States), Siemens (Denmark, merged with Bonus in 2004), and Suzlon (India).

A key development in the global wind power market is the emergence of China as a significant player, both in manufacturing and in the addition of wind power capacity. Five of the largest electrical, aerospace, and power generation equipment companies began to develop wind turbine technology in 2004, and four signed technology-transfer contracts with foreign companies. Such big players are bringing new competencies to the market, including finance, marketing, and production scale, and are adding credibility to the technology. In China, two primary turbine manufacturers, Goldwind and Xi'an Nordex, have market shares of 20 percent and 5 percent, respectively (75 percent of the market being imports). Harbin Electric Machinery Co., one of the biggest producers of electrical generators in China, recently completed design and testing of a 1.2 MW turbine and was working toward production. Harbin's turbine is entirely its own design, for which it has

BOX 3.4

A Case of Other Barriers to Technology Diffusion: The China Study

LAX ENVIRONMENTAL AND REGULATORY REGIMES. The main disincentive to the use of combined cycle for electricity generation in China is that, despite the existence of Chinese regulations, many of which appear to be comparable to those in other countries, there is a widespread lack of enforcement and monitoring. The absence of monitoring means that these regulations have little impact, particularly on the performance of existing coal-fired power plants and industrial installations. Since the enforcement by environmental protection officials is weak, many of these plants do not employ incremental technologies or "end-of-pipe" technologies like electrostatic precipitators or flue gas desulfurization units. If it was otherwise, IGCC would become more attractive.

WEAK INTELLECTUAL PROPERTY RIGHTS (IPR) REGIMES. Regimes for IPR tend to vary widely, especially between developed and developing countries, owing to differing interests, cultures, and administrative capacities. Industrialized countries, which are the main exporters of technologies, tend to see IPRs as a primary means for promoting technology development by offering inventors protection to reap the benefits from their invention. Developing countries are more concerned with having access to existing

claimed full intellectual property rights, the first such instance by a Chinese manufacturer. Dongfang Steam Turbine Works began producing a 1.5 MW turbine and installed four of these in 2005 (REN21 2006).

The industry is also witnessing a rapid globalization of its operations, with many companies considering investments overseas to be competitive. As noted by Brewer (2007), firms sometimes avoid tariffs by undertaking FDI inside the foreign market. Sometimes these projects involve local partners in joint ventures, in which there is the potential for interfirm as well as international technology transfer in both directions. Vestas of Denmark, the leading manufacturer with 30 percent of the global market, opened a blade factory in Australia and planned a factory in China by 2007 to assemble nacelles and hubs. Nordex of Germany began to produce blades in China in 2006. Gamesa of Spain is investing US$30 million to open three new manufacturing facilities in the United States. Gamesa, Acciona of Spain, Suzlon of India, and GE Energy of the United States were all opening new manufacturing facilities in China, with Acciona and Suzlon each investing more than $30 million. The top exporters and importers are presented in table 3.5.

The import-export ratio between high-income OECD and low- and middle-income developing countries between 1995 and 2006 is presented in figure 3.4. The figure suggests that much of the trade has been within developed countries that are still major exporters of wind power technology. Developing countries technologies at affordable costs, and with making them more widely available. Consequently, developing countries tend to have far weaker IPR laws than industrialized countries. Case studies on environmental markets in China (CESTT 2002) mention IPR infringements as a problem, though they are not characterized as a major obstacle.

CASE OF MITSUI BABCOCK. Mitsui Babcock has an extensive presence in China, having won orders for around 5,000 MW of coal-fired utility boilers during the past 20 years, but the company views technology transfer as more of a threat than an opportunity. Unlike competitors such as Combustion Engineering, Mitsui Babcock has not entered into formal licensing agreements or joint ventures with Chinese boiler makers. Instead, the company prefers to work with local Chinese manufacturers on a case-by-case basis. The main reason for this strategy is the mixed experience of Combustion Engineering, which licensed its design to the Ministry of Electric Power. While Combustion Engineering's designs were acquired by all of China's large boiler makers, the resulting licensee revenue has been very small. Instead of following the licensing route, Mitsui Babcock has a wholly owned Chinese trading company (Babcock Shanghai Trading), which has a license to export goods from China and convert local currency into U.S. dollars to generate revenue for the parent company.

Source: Jin and Liu 1999.

TABLE 3.5
Top 10 Trading Countries in Wind Energy

	Exporter	Importer
1	Germany	United States
2	Japan	China
3	Italy	Germany
4	Denmark	United Kingdom
5	Belgium	France
6	United States	Canada
7	Spain	Belgium
8	France	Korea, Rep. of
9	United Kingdom	Italy
10	China	Mexico

Source: WITS database.

have become more active players only in more recent times. The developing countries continue to be net importers, however, on account of their declining import-export ratio; either their level of imports is decreasing, or their level of exports is increasing.

Liberalization of Wind Power Technology

Wind power technology focuses on wind energy generation and is composed of three integral components: gear box, coupling, and wind turbine. The six-digit HS coding system in this analysis closely conforms to the identified technologies.

Even given the rapid growth in wind energy generation, high tariffs are a key barrier with regard to the further expansion of international trade. A sample of maximum tariffs is presented for wind technology for the 18 high-GHG-emitting developing countries. With the exception of three countries—Kazakhstan,

FIGURE 3.4
Wind Power Generation Import-Export Ratio in High-Income versus Low- and Middle-Income Countries

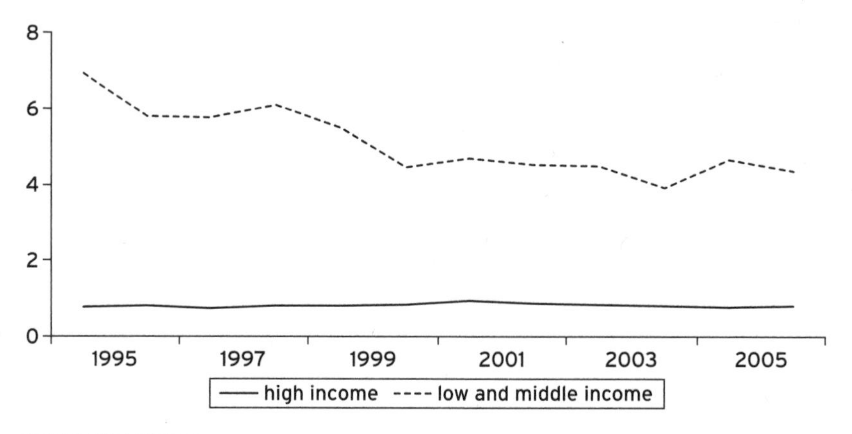

Source: WITS database.

South Africa, and Nigeria—all countries studied institute a tariff varying from 1 percent in Philippines to as high as 15 percent levied by India and some other countries (see table 3.6). When compared with the average industrial tariffs, the tariffs on wind technology are much lower in most countries. For comparison, the table also presents the average industrial tariffs and average tariffs in high-income OECD countries.

In addition, NTBs shown in the table are estimated as ad valorem equivalent and elaborated in percentages that vary across the 18 countries. Seven of the 18 countries levy nontariff barriers, as high as 89 percent by Nigeria and as low as 32 percent by Colombia.

Solar Photovoltaics (PV) Technology

The solar photovoltaics industry is also growing rapidly and is increasingly globalized. In five years, from 1999 to 2004, the solar PV industry quadrupled its cumulative production to more than 4 gigawatts. Production continued to expand aggressively around the world in 2004, and annual production exceeded 1,100 MW. Global production increased from 1,150 MW in 2004 to over 1,700 MW in 2005. Japan was the leader in cell production (830 MW), followed by Europe (470 MW), China (200 MW), and the United States (150 MW).

Market Trends in PV Technology

In China, solar PV cell manufacturing more than tripled, from 65 MW to 200 MW, with manufacturing capacity of about 300 MW by 2005. Module production more

TABLE 3.6

Applied Average Tariffs and NTBs for Wind Technology in 18 High-GHG-Emitting Developing Countries (%)

Countries	Average Tariffs on Wind Technology	Average Industrial Tariffs	NTBs on Wind Technology
Zambia	15	12	60
India	15	29	
Mexico	15	17	
Argentina	14	12	
Brazil	14	14	87
Colombia	10	12	32
Indonesia	10	7	
Thailand	10	16	
Venezuela	10	12	
China	8	10	
Bangladesh	8	18	
Chile	6	6	
Egypt	6	13	70
Malaysia	5	9	59
Philippines	1	6	88
Kazakhstan	0	3	
Nigeria	0	27	89
South Africa	0	8	
High-income OECD countries	3	4	

Source: WITS database.

than doubled, from 100 MW to over 250 MW, with production capacity approaching 400 MW by year-end. Three Chinese PV manufacturers announced plans to expand PV production by more than 1,500 MW by 2008–10 (Nanjing CEEG PV Tech, Yingli Solar, and Suntech Power).

The major global manufacturers are Sharp, Kyocera, and BP Solar, though rapid capacity expansion by many players leads to changes in the top positions year to year. China and other developing countries have emerged as solar PV manufacturers. India has eight cell manufacturers and 14 module manufacturers. India's primary solar PV producer, Tata BP Solar, expanded production capacity from 8 MW in 2001 to 38 MW in 2004. Another 36 MW production line was inaugurated in March 2007 (http://www.tatabpsolar.com). In the Philippines, Sun Power has the capacity to produce 110 MW and is still expanding. Solartron in Thailand announced plans for 20 MW cell production capacity by 2007. Across the whole industry, economies gained from larger production scales, as well as design and process improvements, that promise further cost reductions. The top importers and exporters are presented in table 3.7.

TABLE 3.7
Top 10 Trading Countries in Solar Photovoltaics

	Exporter	Importer
1	Japan	Germany
2	China	United States
3	Germany	China
4	United States	Hong Kong, China
5	Taiwan, China	Japan
6	Malaysia	Korea, Rep. of
7	France	France
8	Korea, Rep. of	United Kingdom
9	Spain	Canada
10	Netherlands	Italy

Source: WITS database.

Figure 3.5 shows the import-export ratio between high-income OECD and low- and middle-income developing countries between 1995 and 2006, which suggests increasing convergence between developed and developing countries. This probably indicates the increasing dominance of some developing countries—such as China—in the global market.

Liberalization of Trade of PV Technology

As PV cells account for more than half of the cost of an installed solar electricity system, reducing tariffs would have a significant effect on overall costs. Maximum import tariffs for the countries examined range from 32 percent to 6 percent, with the exception of Kazakhstan, which has completely liberalized the importing of PV technology. When compared with the average industrial tariffs, the applied tariffs on solar photovoltaic technology are much higher for most countries (table 3.8). For comparison, the table also presents the average industrial tariffs and average tariffs in high-income OECD countries.

FIGURE 3.5
Solar Power Generation Import-Export Ratio in High-Income versus Low- and Middle-Income Countries

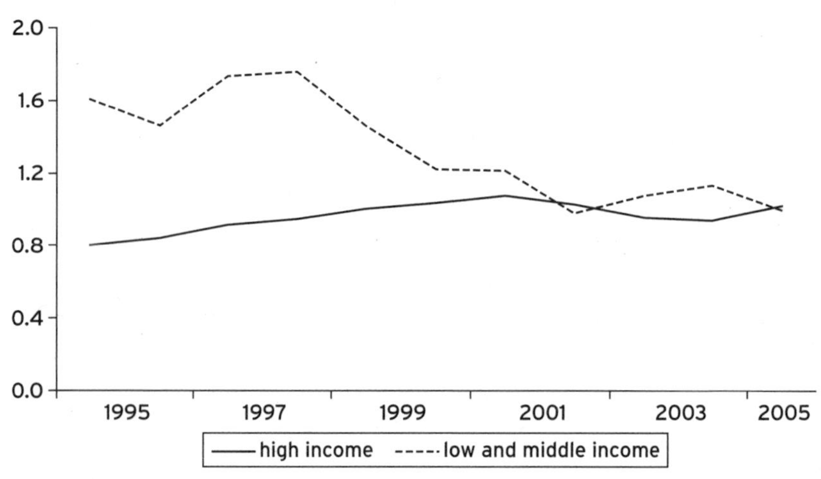

Source: WITS database.

In addition to tariff barriers, of the 18 countries considered in the current study, 5 levy nontariff barriers as high as 70 percent in Nigeria and the Philippines. In addition to these tariff and nontariff barriers, several countries levy other duties that create a huge impediment to the diffusion of technology (box 3.5).

Complementary Policies as Incentives for Renewables

So far, the analysis suggests that in the case of renewables like wind and solar, many developing countries still have high tariff and nontariff barriers. This is, however, just one side of the story. In order to realize the maximum benefits of trade liberalization in clean energy technologies, additional reforms and incentives are required. The recent development of grid-connected renewable energy technologies such as wind demonstrates how countries have influenced the development of clean energy markets through fiscal and financial incentives.

Globally, currently 49 countries have adopted some type of policy target to promote renewable energy power generation. Perhaps the most widespread of these policies are the so-called feed-in laws, such as the well-known PURPA law in the United States, which establish tariffs at which small power producers can

TABLE 3.8

Applied Tariffs and NTBs for Solar Photovoltaic Technology in 18 High-GHG-Emitting Developing Countries (%)

Countries	Average Tariffs on PV	Average Industrial Tariffs	NTBs on PV
Egypt	32	13	
Bangladesh	25	18	
Zambia	30	12	
Nigeria	20	27	70
Argentina	18	12	57
Brazil	18	14	53
Malaysia	18	9	
Colombia	15	12	
Indonesia	15	7	
India	15	29	
Philippines	15	6	70
Venezuela	15	12	
Mexico	13	17	62
South Africa	12	8	
China	10	10	
Thailand	10	16	
Chile	6	6	
Kazakhstan	0	3	
High-income OECD countries	3	4	

Source: WITS database.

BOX 3.5

Cambodia: Additional Duties Leading to Lower Diffusion

In addition to charging tariffs, Cambodia considers PV cells as finished products and charges a 35 percent additional duty. This remains the key barrier to expanding the market for PV products. Analysis suggests that eliminating import duties entirely would cut the cost of purchasing PV systems by 7–10 percent. Eliminating import duties on related components of solar electricity systems—such as storage batteries, charge controllers, compact fluorescent lamps, and inverters—would further reduce costs, making solar PV systems more marketable. Cambodia's renewable energy duties are the highest in ASEAN. In particular, solar photovoltaic goods (e.g., panels, inverters, and controllers) are assessed high duties. Rates for most renewable energy HS classifications are 15 percent. For solar PV systems and related equipment (panels, controllers, inverters, etc.), however, the customs duty is 35 percent (48.5 percent after value added tax).

A related barrier is the lack of detailed specifications of the PV system components in the HS code. Even the more advanced eight-digit AHTN coding system used in ASEAN is inadequate in properly specifying renewable energy and other clean energy technologies. The first step toward easing trade barriers would be to establish clear customs code specifications for renewable energy systems. It must also be recognized that developing an eight- or 10-digit coding system would in itself incur additional administrative expenditures.

Source: Steenblik (2005).

sell power to the utility grid. Initially implemented by the developed economies, they are now widely applied in emerging markets, including several states in Brazil, China, India, Indonesia, Nicaragua, Sri Lanka, and Thailand. Renewable portfolio standards (RPSs), which direct utilities to derive a portion of their total generating capacity from renewable energy, have been adopted not only in the United States and Europe, but also in India and Thailand. The commonly applied fiscal incentives include (i) income tax exemptions, reductions, or credits offering preferential income tax treatment for renewable energy investments; (ii) accelerated depreciation permitting rapid write-off of capital investments in renewable energy equipment; and (iii) a sales tax, VAT, and/or customs duty exemption to reduce the cost of renewable energy investments.

A rich body of experience now exists that relates to the application of financial incentives to promote renewable energy development. Box 3.6 briefly describes the best practices and lessons learned in this area.

BOX 3.6

Lessons Learned in Designing Financial Incentives for Renewable Energy

- *Fiscal incentives work.* Reducing initial investment cost through tax incentives has been proven to stimulate demand for renewable energy in the marketplace in many countries. However, care should be taken to make sure incentives are not offered for products that are already profitable.
- *Fiscal incentives should be temporary.* Fiscal incentives for renewable energy are intended to stimulate less-established technologies and make them more competitive with established alternatives. However, these incentives should not be permanent.
- *Performance-based incentives are effective.* On grid-connected systems, for example, incentives offered on a per-kilowatt-hour basis for delivered power have proved to be more effective than incentives offered against capital costs (e.g., investment subsidies or accelerated depreciation).
- *Fiscal incentives cannot substitute for quality.* Tax and other financial incentives cannot substitute for an adequate infrastructure of, for example, qualified installers and maintenance personnel. First cost is only one of the considerations that cause people to invest in renewable technologies.
- *Burdensome procedures or delay in receiving benefits can nullify incentives.* If the procedures to obtain incentives are complex or time-consuming, they are unlikely to achieve their desired benefit.
- *Fiscal incentives requiring regulatory change and administration are less effective.* By their nature, policies that require tax legislation or regulatory action are slow and may not be able to adapt to change (e.g., technology innovation) in a timely manner.
- *Fiscal incentives may be more effective than petroleum taxes.* Decisions on new investment are typically based on initial investment cost rather than the life-cycle cost of operation.
- *The tax collection system can limit effectiveness of some fiscal incentives.* Tax incentives such as credits and accelerated depreciation have proved effective when tax rates are sufficiently high and the tax collection system is broad based.
- *Fiscal incentive benefits should parallel equipment life-cycle costs.* There have been cases in the past where financial incentives to developers have greatly exceeded the cost of development, resulting in unnecessary incentive costs and sometimes overcapacity. One clear advantage of import duty reduction or exemption is that the size of the incentive is directly linked to the scale of the project investment.
- *Fiscal incentives are not a "quick fix" solution.* These incentives are only one of many policy tools, and they should be used to complement other policy initiatives.

Source: Vernstrom 2007.

TABLE 3.9
Top 10 Trading Countries for Fluorescent Lamps

	Exporter	Importer
1	China	United States
2	Hungary	France
3	Poland	Germany
4	Netherlands	United Kingdom
5	France	Italy
6	Canada	China
7	Indonesia	Netherlands
8	United States	Japan
9	Italy	Canada
10	Japan	South Africa

Source: WITS database.

Energy-Efficient Lighting

The efficacy of lighting systems vary significantly from sector to sector, ranging from as low as 20 lumens per watt (lm/W) in the residential sector to as high as 80 lm/W in the industrial sector. From a technological perspective, the low efficiency achieved in the residential sector is, to a large extent, due to the important role of incandescent lamps, which are characterized by very low energy efficiency.

Substituting incandescent lamps with fluorescent lamps can therefore be an effective means to improve residential sector lighting efficiency, as they consume only 20 to 25 percent of the energy that incandescent light bulbs use to provide the same level of light. While fluorescent lamps have a higher initial cost, due to their low energy use, on a life-cycle basis they are significantly more economical than incandescent lamps.

In terms of the major trading countries, it is interesting to note that, among developing countries, China and Indonesia have emerged as major players in the florescent lamps market (table 3.9). Examining the import-export trends suggests that the market is growing very rapidly for florescent in developed countries, and much of the supply is coming from China and other developing countries (figure 3.6). These ratios suggest a role for multilateral liberalization of energy-efficient lighting.

Liberalization of Trade in Fluorescent Lamps

This section assesses the existing tariff and nontariff barriers on efficient lighting based on the proxy of fluorescent lamps in 18 developing countries selected from a list of non–Annex I countries. On assessing the tariff levels across 18 countries, the study finds that, with the exception of Kazakhstan, all countries have maximum levels of tariffs varying from 5 percent to 30 percent. In the data analyzed, the tariff on fluorescent lamps is the highest across all other clean technologies assessed in the data. The highest tariffs on fluorescent lamps are applied by Malaysia and Zambia. For most countries, the applied tariffs on fluorescent lamps are much higher than the average industrial tariffs (table 3.10).

For comparison, the table also presents the average industrial tariffs and average tariffs in high-income OECD countries. The NTBs in the form of quotas and import ceilings applied across 18 countries are translated into percentages. As the table

FIGURE 3.6
Import-Export Ratio of Fluorescent Lamps in High- and Low-Income Countries

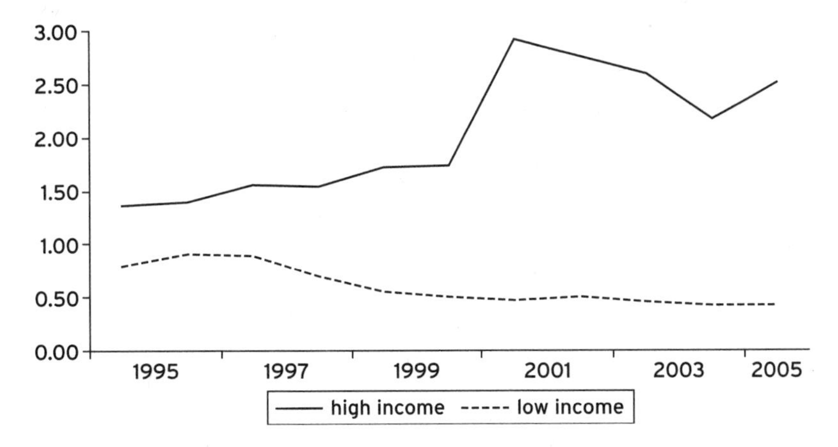

Source: WITS database.

TABLE 3.10
Average Applied Tariffs and NTBs on Fluorescent Lamps in 18 High-GHG-Emitting Developing Countries

Country	Average Tariffs on Fluorescent Lamps	Average Industrial Tariffs	NTBs on Fluorescent Lamps
Zambia	30	12	83
Malaysia	30	9	85
Colombia	20	12	
Nigeria	20	27	91
Thailand	20	16	
Venezuela	20	12	
Bangladesh	19	18	
Argentina	18	12	
Brazil	18	14	96
Egypt	18	13	87
South Africa	17	8	
India	15	29	102
Mexico	15	17	
Philippines	11	16	93
China	8	10	
Chile	6	6	
Indonesia	5	7	
Kazakhstan	0	3	
High-income OECD countries	4	4	

Source: WITS database.

shows for fluorescent lighting, seven of the 18 countries levy NTBs for fluorescent lighting that range from as high as 102 percent in India to 82 percent in Algeria. For all other countries, NTBs are almost nonexistent.

Compact fluorescent lamps (CFLs) which are classified as a part of the broader florescent lamps HS category especially offer a win-win-win alternative, with climate, economic, and—to the extent that their use displaces the consumption of risk-prone fossil fuels and reduces system load—energy security benefits.[5] Yet CFLs account for only 6 percent of the lighting market and represent a minor share of light production in the residential sector. The natural uptake of CFLs in the market is hampered by a variety of barriers. Though CFL costs have gone down significantly since they were first introduced, their high initial cost compared with incandescent lamps remains an important barrier, particularly for the poorer sections of the community.

Policy Mix for Extending Liberalization Potential in Efficient Lighting

A country wanting to set up a policy program to enhance CFL diffusion has to clearly identify barriers specific to its socioeconomic circumstances in order to optimize policy choices. Evidence shows that the most successful cases are conditional, where policy addresses multiple barriers. First, the cost and information barriers need to be addressed jointly. Second, governments should consider a portfolio approach, with different measures targeting different barriers. Box 3.7 highlights a combination of policy measures employed by South Africa that led to an increased absorption of compact fluorescent lamps in the residential sector and corresponding energy savings from their use.

Conclusions

Liberalizing trade and identifiable nontariff barriers in clean energy technologies could result in gains in the volume of trade. As shown in the analysis for the four technologies, these gains compare to a 7 percent increase in trade volume when the only tariff barriers are eliminated without a change in nontariff barriers across the sample countries. As the results also suggest, the net effect could vary across technologies and across countries, depending on the existing barriers and the import elasticities of demand.

This finding has important implications for GHG emissions, given that even within a small subset of clean energy technologies and a select group of countries, impacts of trade liberalization can be reasonably substantial. In addition, these technologies will also confer local environmental benefits and general efficiency improvements in the production process. Further, liberalizing trade in renewables will change the Clean Development Mechanism baseline for renewable energy projects and could thereby facilitate more high-end and state-of-the-art

BOX 3.7

Bundling Policies to Promote Energy Savings: The Case of South Africa

South Africa was one of seven countries to take part in the World Bank/GEF Efficient Lighting Initiative (ELI). Bonesa—the company set up to operate ELI in South Africa—focused on 50,000 houses and aimed at replacing all lamps with CFLs. The initiative was successful in reducing the price of CFLs from R 60–80 per lamp in 1998, to R 13–20 in 2004, according to Eskom, the state-owned national utility. By training staff from the local communities, Bonesa also contributed to enhancing local expertise in CFL technologies and advantages. At the end of the three-year period, Eskom implemented a residential demand-side management (DSM) program.

The program had a twofold objective: (i) to reduce electricity demand at peak periods by shifting load to off-peak hours, and (ii) to reduce overall electricity demand through the implementation of energy efficiency measures. The national efficient lighting roll-out initiative, part of the DSM program, was launched by Eskom in 2002 to provide lower-cost alternatives by focusing on the effective use of electricity. Between 2003 and 2005, about 2.5 million CFLs were distributed at subsidized prices. Similar initiatives were later reproduced through regional programs, which further helped the replacement of over 4.3 million CFLs within three months (Eskom 2005), mainly through door-to-door giveaway campaigns. This led to a 193 MW savings, exceeding the 155 MW target.

The DSM program also launched awareness campaigns, which sought to familiarize consumers with the environmental and financial benefits of CFLs. "Power alert" campaigns were also launched to limit peak demand: consumers were encouraged to stay tuned to their radio stations and televisions to hear suppliers communicate the level of shortage. With a four-color code ranging from green to brown, Eskom intends to alert consumers of the implications in their use of electricity and appliances in the household. By the end of the ELI program, CFL sales had increased by 64 percent, while the sales of incandescent lamps decreased by 9 percent (GEF 2004). The initiative contributed to lower CFL prices, raised awareness, and enhanced awareness of CFL benefits.

Source: Lefèvre and others 2006.

technology transfer through CDM projects. The analysis does not take into account other barriers of IPRs, government distortionary policies, and investment barriers, which, as seen through specific cases, still impede technology diffusion. Despite these impediments, developing countries are playing a larger role in technology development and exports.

Key Findings from Chapter 3

- A number of low-carbon technologies exist that can be effectively utilized to combat climate change.
- Levels of tariffs and nontariff barriers for clean energy technology vary across the 18 high-GHG-emitting developing countries identified for the trade analysis.
- The elimination of tariffs and NTBs could lead to a considerable increment in the volume of clean energy technologies traded, from 7 percent in the case of tariff removal to 14 percent for removal of tariffs and NTBs. This result is evident across the four technologies identified for the study: clean coal, wind, solar, and energy-efficient lighting.
- In addition to tariffs and NTBs, a country's investment climate and intellectual property rights regime significantly influence technology diffusion.
- The scope for South-South and North-South trade cannot be underestimated, given the large, evolving export potentials of developing countries.

Notes

1 According to the IEA, fuel combustion (production, transportation, and consumption) is responsible for the largest share of global anthropogenic greenhouse gas emissions, accounting for about 80 percent of greenhouse gas emissions (IEA Database 2006).

2 Global emission scenarios will look quite different if we include emissions from deforestation and land use changes. Consideration of those issues is beyond the scope of this study.

3 Recommendations of the Expert Group on Technology Transfer for enhancing the implementation of the framework for meaningful and effective actions to enhance the implementation of Article 4, paragraph 5, of the Convention. FCCC/SBSTA/2006/INF.4.

4 These countries are Argentina, Bangladesh, Brazil, Chile, China, Colombia, Egypt, India, Indonesia, Kazakhstan, Malaysia, Mexico, Nigeria, Philippines, South Africa, Thailand, Venezuela, and Zambia.

5 CFLS are classified at HS eight-digit level, thereby making it difficult to track their volume of trade across countries.

CHAPTER 4

Opportunities for Win-Win-Win: Liberalizing Trade in Environmental Goods and Services

The analysis thus far suggests that access to clean energy technology is essential, especially for developing countries seeking to diversify their energy sources and to reduce carbon emissions without hindering economic development. This would entail not only removing trade and investment barriers that inhibit cleaner-technology diffusion, but also putting in place regulatory and market mechanisms needed to assure investors and researchers that a market for new technologies will exist.

Coincidentally, the broader subject of liberalizing trade in environmental goods and services (EGS) is on the agenda for the first time in the WTO's Doha Round of negotiations.[1] Currently the negotiations, centered on the identification of environmental goods to be liberalized, are taking place in special sessions of the WTO Committee on Trade and Environment. Environmental services are being negotiated separately with the Council for Trade in Services. While the mandate does not explicitly exclude consideration of agricultural products as environmental goods, for example, ethanol, so far WTO members have formally tabled only industrial products for liberalization purposes.

Many experts believe that by singling out within a special mandate those goods and services that previously would have been negotiated as a matter of course, trade negotiators consider the liberalization of these goods particularly significant. It is widely accepted that trade liberalization of EGS would benefit the environment by contributing to lowering the costs of goods and services necessary for environmental protection, including those beneficial for climate change. By enabling a level playing field in terms of conditions of competition between climate-friendly technologies, goods, and services, liberalization could also contribute to increasing the range of choices available to countries to tackle challenges related to climate change.

The chapter is organized as follows. The first section provides an overview of the current debate on environmental goods within the WTO, highlighting the key issues and challenges for negotiation and will attempt to examine each of these in terms of their significance for climate (change objectives. The second section discusses the course of negotiations in terms of their relevance for the purposes of climate change mitigation. The next section explores the options for negotiating a separate climate-friendly package within EGS. The fourth section discusses issues and challenges relevant for negotiating a climate-friendly package. The final section concludes with what could be some of the essential ingredients of this climate-friendly package.

It is pertinent to note here that environmental services (ES) negotiations are also relevant for climate change mitigation. Arguably, a number of services, such as protection of ambient air and climate (through the cleaning of exhaust gases and carbon capture and sequestration) directly address climate change.

However, the modes of trade are different from those for goods (often involving investment, or "commercial presence"). While there are issues with regard to classification, the definitional complexity is certainly less than for environmental goods (EG), and the trade issues are of a different nature and often involve domestic regulatory issues as well. Thus, this chapter focuses only on EG negotiations. Suffice it to say that since EG and ES are very often supplied as an integrated package, WTO members will need to be mindful of the synergies between goods and services and ensure coherence between both negotiations accordingly.

Complexity Surrounding Environmental Goods Discussions

While support for environmental technology can provide a "win-win-win" example of trade liberalization benefiting the environment, the debate at the WTO has been far more complex and has thrown up a number of issues that need to be addressed if the negotiations are to have a meaningful outcome for sustainable development. These challenges highlight the various political economy concerns stemming from domestic economic and social considerations, as well as the manner in which products are classified for the purpose of international trade.

The debate includes varied perceptions on the definitions, reach, and range of technological options. The main concern, notably for developing countries, is *what* products are included and *how* they are liberalized. These interpretations of

BOX 4.1

Main Issues in Liberalization of Environmental Goods and Services

All WTO members agree that liberalization of environmental goods should be geared toward environmental protection. The fundamental fault lines of disagreement are underpinned by different perceptions of what "environmental goods" are (i.e., the issue of definition), which would determine what goods to include or not for liberalization under the mandate and how to liberalize in a manner that addresses the interests of both developed as well as developing countries (i.e., the issue of approaches to liberalization).

The key issues surrounding what to liberalize include (i) dealing with single versus dual-use goods; (ii) the relative environmental friendliness of goods; (iii) dealing with the constantly evolving technology; (iv) assessing implications for domestic industries, especially in developing countries; (v) dealing with nontariff barriers; (vi) enhancing opportunities for developing country exports; and (vii) dealing with agricultural environmental issues.

The key issues surrounding how to liberalize have been divided between the "list" approach and the "project" approach. Developed countries interested in liberalizing environmental goods support a list approach; that is, focusing on identifying and submitting specific lists of goods and then negotiating the elimination or reduction of bound tariffs (and nontariff barriers) permanently and on a most-favored-nation (MFN) basis. Developing countries such as India prefer a project approach; that is, liberalization would be bound temporally and only for the duration of environmental projects that would benefit from liberalized imports of goods and services on an MFN basis. This would be approved by a designated national authority (DNA) based on criteria developed by the Committee on Trade and Environment (CTE).

definitions and roles can perceptibly change the sustainable development impacts arising from trade liberalization, specifically for developing countries. Box 4.1 provides a quick overview of key issues that have surfaced during the course of negotiations. These are spelled out in some detail in the later sections of the chapter.

Traditional Environmental Goods (Environmental End Use) and Environmentally Preferable Products

Despite the lack of a universally accepted definition for EGs, environmental goods could be conceptualized in two ways. The first is the narrow, conventional conception that focuses on treating a specific environmental problem through the end use of a particular good or service. This characterizes the traditional classification of EGs and includes goods such as wastewater treatment or air-pollution-control equipment that have an environmental end use; that is, they directly address an environmental problem.

The second conceptualization is broader and includes environmentally preferable products (EPPs). The United Nations Conference on Trade and Development defines EPPs as products that cause significantly less "environmental harm" at some stage of their life cycle than alternative products that serve the same purpose, or products whose production and sale contribute significantly to the preservation of the environment (UNCTAD 1995). In this case, the primary purpose of the product or service is usually not to remedy an environmental problem. The environmental benefits may arise during and as a result of the production method, during the course of its use, or during the disposal stage of the product (Sugathan and others 2007). A wide array of products—ranging from hybrid cars to energy-efficient washing machines to ethanol—could all conceivably be classified as EPPs.

Most WTO members have sought to avoid including products that were deemed environmentally preferable based on their process and production methods (PPMs) (figure 4.1). This implies, for instance, that aluminum produced using renewable energy as an input is not likely to be included as an "environmental good," since customs authorities would find it physically indistinguishable from aluminum produced through coal-generated electricity. Differentiating goods on the basis of production would also potentially throw up challenges with regard to classification of products for the purpose of trade under the existing and widely used

FIGURE 4.1
Traditional Environmental Goods versus Environmentally Preferable Products

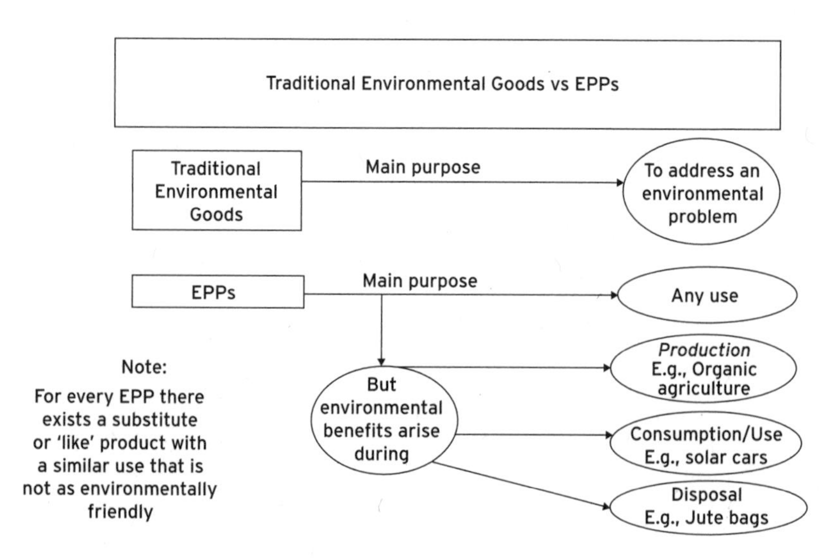

Source: Sugathan and others 2007.

Harmonized Commodity Coding and Description System (HS), as well as require labeling or certification introducing additional complexities of standard-setting, recognition, and acceptance.

Single- versus Dual-Use Goods

A fundamental fault line characterizing these negotiations has been the issue of what constitutes an environmental good. WTO members did not start the negotiations by attempting to define what environmental goods are; instead, some countries adopted an approach of drawing up lists of products that they considered important for environmental protection purposes. A number of these products were derived from a list drawn up by the Asia Pacific Economic Community (APEC) when it selected the environmental goods sector for inclusion in its Early Voluntary Sector Liberalization initiative launched in 1997. The APEC list, drawn up on the basis of individual nominations, owes its origins to the OECD/Eurostat (Statistical Office for the European Communities) definition of the environment industry that was developed for analytical purposes. The industry, according to the OECD and Eurostat, comprises "activities which produce goods and services to measure, prevent, limit, minimise or correct environmental damage to water, air and soil, as well as problems related to waste, noise and ecosystems." The OECD has categorized these goods and services under three broad headings: pollution management, cleaner technologies and products, and resource management. Experts have, however, pointed out "inclusion and exclusion" differences in the listing categories of goods. For example, ethanol is included in the OECD list, but excluded from the APEC list.

Despite the fact that members sought to avoid defining an environmental good, the issue resurfaced when it was pointed out that the majority of the products submitted by interested members were those that had dual uses; that is, they were used for both environmental and nonenvironmental purposes. Two types of dual-use products can be identified. The first comprised products that intrinsically had more than one use. A good example is a pipe, which can be used as an input to a renewable energy plant or wastewater treatment plant but can also be used to transport oil. Should a pipe therefore be liberalized as an environmental good?

The second type of dual-use issue characterizes products classified under the HS, which as mentioned before is a product classification system widely used in international trade. As also noted in chapter 3, the challenge is that WTO members have HS codes for product categories only up to the six-digit level. For the purpose of tariff liberalization at the WTO and subsequent implementation by customs worldwide, it is therefore easier to liberalize the whole HS six-digit category, rather than try to isolate and liberalize specific products for which no uniform code exists (also known as "ex-outs") beyond six digits.

List, Project, and Integrated Approaches to Liberalization

The question of approaches to liberalizing environmental goods has proved a major cause of deadlock in the negotiations. Developed countries interested in liberalizing environmental goods adopted a list approach to negotiations on environmental goods similar to that adopted for industrial goods; that is, they identified and submitted lists of specific goods and then negotiated the elimination or reduction of bound tariffs (and nontariff barriers) permanently and on a most-favored-nation (MFN) basis.

However, during the course of the negotiations, developing countries such as India proposed an alternative approach to liberalization that was termed the *project* approach (TN/TE/W/51, TN/TE/54, TN/TE/60 and TN/TE/W/67). The project approach, first proposed in June 2005, was driven by concerns that, owing to complexities surrounding HS codes, the administration of commodities whose classification extends beyond HS six-digit level and the dual uses of a majority of environmental goods, the list approach would lead to far greater liberalization than intended, extending to goods with both environmental and nonenvironmental end uses. In the project approach, the environmental projects that would benefit from liberalized imports of goods and services would be approved by a designated national authority based on criteria developed by the WTO Committee on Trade and Environment (CTE). Further, domestic implementation of these criteria would be subject to WTO dispute settlement.

According to some WTO members, the project approach has two drawbacks. It lacks binding and predictable market access offered on a permanent basis, and it is inconsistent with WTO rules. Concerns have also been raised regarding the time taken to develop multilateral criteria as well as time needed for dispute-settlement proceedings relative to the duration of a project (Sugathan and others 2007).

Argentina proposed an integrated approach (TN/TE/W/62) in October 2005. Under that approach, national authorities would decide whether or not to temporarily eliminate tariffs for environmental products used in particular environmental projects. Members within the CTE special sessions would multilaterally preidentify categories of environmental projects and environmental goods that could be used in them. However, proponents of the list approach still consider the integrated approach inadequate, as it did not meet the criteria of binding and predictable market access and consistency with WTO rules.

Linking of Current EG Discussions to Climate Change Mitigation

From the perspective of developing countries, the choice of a negotiating approach needs to be resolved before discussions on specific products take place. This has contributed to a negotiating deadlock, as supporters of the list approach and the project approach have so far refused to compromise on their respective approaches.

A recent informal submission by Colombia in June 2006 (JOB(06)149) attempts to bridge the various approaches. This approach would require members to define clear criteria for a single environmental end use, namely, improving the environment or reducing waste and the consumption of natural resources, and having a direct and verifiable environmental application that complies with the objectives of multilateral environmental agreements (MEAs). Products with dual uses would be liberalized if they were used in a project, program, plan, or system deemed to have verifiable environmental benefits by a designated national authority. If this approach is accepted, it would also be applicable to discussions about clean energy technologies for climate change mitigation (ICTSD 2006).

Categorizing of Climate-Friendly Goods for Trade Negotiations

The recent report on climate change mitigation measures issued by the IPCC in Bangkok states that the technologies with the largest economic potential for the respective sectors include energy supply, transport, buildings, industry, agriculture, forestry, and waste. It also noted that energy efficiency "plays a key role across many scenarios for most regions and timescales" (IPCC 2007).

A list of 153 environmental goods was submitted as an informal document (JOB(07)/54) in April 2007 by the Friends of EGS Group, comprising Canada, the EU, Japan, Korea, New Zealand, Norway, Switzerland, Chinese Taipei, and the United States, for discussion in the WTO. The Friends Group, however, retains the discretion to add more products to the list in the future. From this list, the study identified about 40 goods that can be broadly categorized as climate friendly. This goes beyond the technologies discussed in chapter 3. The study identified global trade trends in the technologies, along with the existing tariff barriers. This analysis examines and suggests a narrower choice of climate-friendly products that would be acceptable to a broader range of countries, rather than a broader range of goods that would be acceptable to only a few countries.[2]

The analysis of the 40 identified climate-friendly technology products suggests that their use has seen a considerable increase in the past few years, with the global trade almost doubling (from US$67 billion to US$119 billion) since 2002. The trade data for high-income and low- and middle-income WTO members, examined separately, suggests that even in the low- and middle-income countries, the trade in climate-friendly technologies is growing rapidly, though these countries continue to be net importers overall (table 4.1).

The current list does not include other low-carbon technologies such as biofuels, which also have potential for climate mitigation. However, including some of them may involve agricultural trade liberalization (e.g., ethanol is considered an agricultural good), which has proved more controversial under the WTO.

An examination of existing trade barriers for these technologies suggests that among low- and middle-income WTO members, considerable barriers to entry

TABLE 4.1
Trade in Climate-Friendly Technologies of Both High-Income and Low- and Middle-Income WTO Members

	High-Income WTO Members US$ 000s		Low- and Middle-Income WTO Members US$ 000s	
Year	Imports	Exports	Imports	Exports
2002	24,865,316	26,629,191	14,650,587	9,229,445
2003	27,605,322	29,677,598	17,649,253	10,951,796
2004	35,513,734	40,212,179	23,847,009	14,784,814
2005	42,023,036	46,087,645	27,318,520	18,605,985

Source: WITS database.

still exist for these technologies (see appendix 6). Both maximum-bound and applied-average tariffs continue to remain much higher than those in the high-income WTO member countries. The opportunity that exists for lowering trade barriers would serve both trade and climate change interests. Since low- and middle-income developing countries are also emerging as major importers as well as suppliers of these commodities, it would also be in their interest to bargain for reducing or eliminating the restrictions.

Options for Negotiating a Climate-Friendly Package within the WTO Framework

The previous sections highlighted some of the main challenges within EG negotiations, as well as the relevant goods that are important from a climate change perspective. It is beyond the scope of this study to recommend whether a list, project, or integrated approach would be the most appropriate method to follow. While the methodology under the project approach is straightforward, the integrated approach and the list approach would still involve the exercise of identifying relevant environmental goods at the multilateral level.

It is highly likely that the current negotiations surrounding EGS will be a long, drawn-out process. Recognizing the underlying challenges and difficulties of reaching agreement on the various contentious issues, we believe it would be useful to learn from past rapid liberalization initiatives, most notably the Information Technology Agreement. Another option for resolving EG negotiations is to consider a plurilateral agreement along the lines of the Agreement on Government Procurement (GPA), which would be outside the single undertaking (whereby members representing a minimum percentage of trade in climate-friendly products could join), with trade benefits extending only to signatories to the agreement.[3]

In both cases, the package could represent a more ambitious subset of products deriving from the larger environmental goods negotiations, with the aim of immediate elimination of tariffs and, subsequently, nontariff barriers.

Experience from Liberalization Initiatives in Information Technology

The Ministerial Declaration on Trade in Information Technology products, also known as the Information Technology Agreement (ITA), was concluded at the Singapore Ministerial Conference in 1996. Initially, the agreement was signed by 29 members of the ITA (including the then EU-15 as one entity). The agreement did not immediately take effect owing to the requirement that participants representing at least 90 percent of world trade in these products would have to notify their acceptance of the agreement by April 1, 1997. The signatories met the deadline; and the agreement came into effect, with binding first-phase reductions in tariffs by July 1, 1997, followed by second and third rate reductions (by January 1, 1998, and January 1, 1999, respectively), and finally complete elimination of customs duties no later than January 1, 2000.[4] For the majority of developing countries, the end date for phaseout was 2005. At present, more than 70 countries and customs territories are in the process of acceding to the ITA (43 ITA members, counting the EU-27 as one entity). Members of the agreement represent 97 percent of world trade in IT products.[5]

The ITA is essentially a tariff-cutting mechanism. While the declaration provides for a review of NTBs, there are no binding commitments on nontariff barriers. All participants have to comply with three basic principles of the ITA: (i) all products listed in the declaration must be covered, (ii) tariffs on all products must be reduced to zero, and (iii) all other duties and charges must be bound at zero. While there are no exceptions to product coverage, an extended implementation period for sensitive items is possible. Significantly, commitments undertaken in the ITA are on an MFN basis, with benefits extending to all other WTO members.[6]

How ITA Was Achieved: Getting the "Atmospherics" Right

In exploring the possibilities for an agreement on climate-friendly goods to be a concrete deliverable of the broader environmental goods negotiations, it is important to understand the political economy considerations and the factors that led to an economically and politically conducive environment. In other words, the right "atmospherics" for trade liberalization initiatives in information technology eventually led to the ITA.

There were two key factors in getting the atmospherics right. First, there was the realization among all countries concerned about the growing importance of IT industries in terms of employment, innovation, technology diffusion, skills upgrading, foreign direct investment, capital formation or exports, and the explosive growth in IT trade. Second, there was a growing appreciation of the value of IT products as important intermediates in production, with consequent economy-wide benefits. These two factors, together with consensus among a critical mass of IT-producing nations regarding the benefits of tariff-free trade in IT products, contributed to setting the stage.

However, negotiations leading up to the ITA were far from easy. Similar to negotiations on environmental goods, the negotiations for the ITA were also affected by differences over product coverage, with different countries seeking to exempt certain products. The IT sector (similar to the environmental goods) was one in which a number of developing countries were not significant suppliers of products, which industrial countries had listed as a priority for tariff cuts. Several countries—including Korea (Rep. of), Taiwan (China), and Hong Kong (China)—made it clear that negotiating modalities had to be framed in such a way as to be acceptable to the less-developed member countries of APEC.

Under pressure to broaden country coverage to be more responsive to developing-country concerns, the United States and the EU signaled greater flexibility on product coverage and implementation of tariff cuts, although they made it clear that there were limitations on "special treatment" to countries that were competitive producers. A compromise on disputed product coverage between the United States and the EU in the 1996 Singapore Ministerial Conference was another essential element in making a breakthrough possible. The ITA's built-in mechanism for periodic review may have tempered the disappointment among many countries over the initial exclusion of certain items, such as consumer electronics. Later on, flexibility in terms of longer implementation periods was granted to developing countries, as part of efforts to accommodate more members in order to reach the critical threshold of 90 percent coverage of world trade in IT products.

Continuing Challenges for the ITA

The ITA has been regarded as a major success since the WTO's establishment due to the ambitious tariff liberalization initiative involving major developed and developing countries. In a recent speech marking the 10th anniversary of the ITA, WTO Director General Pascal Lamy noted that world exports of ITA products had more than doubled since 1997 in dollar terms, reaching US$1,450 billion in 2005, with annual average growth of 8.5 percent (WTO 2007).

However, implementation of the ITA has run into challenges that hold important lessons with regard to the design of any future climate technology agreement. The Ministerial Declaration and Implementation Documents for the ITA provide for review of product coverage every three years,[7] but no new products have been added since 1996. Despite additional products being submitted for inclusion by a few countries, the review process continues to be a stalemate.

Lessons from Current EG Discussions for Negotiating a Climate-Friendly Package

As clearly revealed in the analysis of issues and challenges confronting WTO negotiators as well as experience from the ITA negotiating process, it will not be an easy task to forge a similar agreement for climate-friendly technologies. However, certain lessons emerge from the previous analysis that could comprise preconditions for (and inform) the design of any such agreement. These are outlined below.

For identifying and liberalizing climate-friendly goods and technologies, the single- and dual-use concern is significant. While it may sound logical to propose liberalization of both type of goods in the interests of cost-effective climate mitigation, in reality this approach may face challenges owing to concerns such as the impact of broader liberalization on domestic industries and jobs and, in some cases, on tariff revenues. The reluctance to pursue more ambitious liberalization may also stem from strategic considerations resulting from the lack of meaningful progress in other negotiations of interest to parties, such as agriculture. If a meaningful climate-friendly package of goods is to be drawn up, these concerns will need to be addressed through appropriate rules and flexibilities, as the package may inevitably include many dual-use products and ex-outs.

Relativity of Environmental Friendliness

In the wide array of climate-friendly goods, many are environmentally friendly. However, there may be substitutes that are less or more preferable environmentally. So how should environmental goods negotiations treat these products? A good example from the ongoing negotiations is natural gas and natural gas–related technologies proposed by Qatar in its submissions (TN/TE/W/14, 19, and 21) as a bridge to a carbon-free era. Qatar also pointed to its role as a backup for wind and photovoltaic systems and a source of manufacturing hydrogen.

In its submission, Qatar rightly maintains that natural gas has been recognized in the Kyoto Protocol negotiations as part of the solution to stabilize greenhouse gases in the atmosphere. The IPCC assessment reports have also recommended increased use of natural gas over other fossil fuels as a way to reduce greenhouse gas emissions.

While natural gas is certainly a cleaner alternative to coal, it is less so compared to ethanol, wind power, or hydrogen. However, the latter may require some amount of subsidies to be viable. Therefore, should such subsidies be removed if they act as de facto nontariff barriers to natural gas? Further, in the interests of meaningful climate change mitigation, removal of trade barriers to natural gas must be accompanied by removal of subsidies to coal that are prevalent in both OECD and non-OECD countries and are a greater threat to global warming (table 4.2). However, once trade barriers are lowered and bound on "relatively friendly" goods and technologies, it may not be possible to raise them again.

Some experts believe that the decisions made with respect to designing specific EGS for trade liberalization will affect the options that shape future R&D decisions among producers of both agricultural and manufacturing goods (Mytelka 2007). It may be better to provide an immediate trade-preference course to clean technologies, such as for goods producing zero emissions and standards that are

TABLE 4.2
Fuel Subsidies in OECD and non-OECD Countries (US$ billions)

	OECD Countries	Non-OECD Countries	Total
Coal	30	23	53
Oil	19	33	52
Gas	8	38	46
Fossil fuels	57	94	151
Electricity[a]	–	48	48
Nuclear	16	Negligible	16
Renewables	9	Negligible	
Nonpayments and bail-out[b]		20	20
Total	82	162	244
Per capita (US$)	88	35	44

a. Subsidies for electricity countries have been attributed to fossil fuels according to the shares.

b. Subsidies from nonpayments and bail-out operations have not been attributed to energy sources.

Sources: Van Beers and de Moor (1998) and IEA (1999).

easily measurable, rather than for goods and technologies that are relatively clean. The advantage of including clean and renewable technologies, but not "relatively cleaner" products, is that it provides a trade incentive for innovation into the former category and would go beyond short-term considerations.

Dealing with Evolving Technologies and Products

During the course of negotiations, Japan introduced a number of energy-efficient products such as washing machines and dishwashers, as well as hybrid cars, as clean technology products. This has also created some controversy, because these products would entail the creation of a separate tariff category for energy-efficient products within the HS classification. Further, energy efficiency is an evolving concept dependent on technology. What happens if a superior substitute evolves in the future or technology embedded within a product becomes better? In cases where the HS code stays the same despite the change in technology, it may not affect the trade concession granted to the product through the negotiations. But at other times, new or superior products may arise that may need a new HS code or revision of existing HS codes.

Some countries, such as New Zealand (TN/TE/W/49), have proposed the creation of a "living list," given the dynamic and evolving nature of the environmental goods sector and the fact that it is continually developing in new and often unexpected directions. The submission by Canada, the EU, Japan, Korea (Rep. of), New Zealand, Norway, Chinese Taipei, Switzerland, and the United States also provides for the development of a review mechanism for any agreed-upon set of environmental goods to take into account dynamic changes.[8]

Many developing countries, however, have concerns about a living list and the implication of automatic liberalization of new products and technologies that such a list might imply. Environmental goods relevant to climate change mitigation will also be affected by technological change and evolution of new products, and trade negotiators will need to respond to such changes if the intention is to maintain zero or a low level of trade barriers for the latest technologies. As in the case of the "relatively friendly" products described above, raising tariffs on older and less environmentally friendly products and technologies may be difficult once these have been lowered and bound.

The Impact of Liberalization on Domestic Industries

A number of developing countries are concerned about the impact of liberalization on existing domestic industries and in some cases on tariff revenue. Some countries, such as China, have proposed a "common" list that would include environmental goods of export and import interest to developing countries. It further proposes a "development list" that would be derived from the common list and comprise goods eligible for special and differential treatment in the form of lower levels of reduction commitments for developing countries (TN/TE/W/42). A recent submission by Canada, the EU, Japan, Korea (Rep. of), New Zealand, Norway, Chinese Taipei, Switzerland, and the United States provides for elimination of tariffs "as soon as possible, but no later than 2008 for developed countries, and those developing countries declaring themselves in a position to do so. For other countries tariffs should be eliminated by X years thereafter." The submission also welcomes specific suggestions from other members about how to implement special and differential treatment for these negotiations.

Concerns about the impact of EG liberalization on domestic industries may be relevant even to goods that are beneficial for climate change mitigation. Many developed countries and developing countries such as China and India have domestic industries engaged in the manufacture of goods such as solar panels and wind turbines (often with the aid of domestic subsidies). Any package for liberalizing climate-friendly goods may also need to respond to these concerns and take into account subsidies and other measures put in place by governments to encourage the domestic renewable energy sector.

Enhancing Export Opportunities for Developing Countries

An earlier analysis of the environmental goods in the APEC and OECD lists (WTO 2002) indicates that developing countries on the whole are net importers of environmental goods, with exports primarily oriented toward regional markets. The balance of trade for developing countries appears to be improving (UNCTAD 2003). The developed world—notably the EU, United States, and Japan—have considerable surpluses in trade in environmental goods (Vikhlyaev 2003).

Developing countries could therefore perceive the environmental goods negotiations as focusing primarily on products of export interest to developed countries and would like to see the inclusion of more products of export interest to them. However, with the exception of Chinese Taipei, Korea (Rep. of), and Qatar, no developing country at the time of writing has formally put exports on the table, though countries like Brazil have referred to the possibility of including ethanol. A number of EPPs of export interest to developing countries—such as organic fertilizers, jute, sisal, and other textile fibers—have been proposed by Switzerland, the United States, and New Zealand.

While organic agricultural products were alluded to by Kenya and other African countries early on in the negotiations (TN/MA/W/40), there was no formal push for including them, owing to reluctance among WTO members to include products based on the PPM criteria. Cuba has also proposed, as a form of special and differential treatment, low enough tariffs on developing-country EG exports in developed-country markets to permit effective entry and approval, mutual recognition, and financial and technological support measures to achieve such entry where the goods are subject to nontariff barriers (TN/TE/W/69). However, some developing countries may have dynamic-export interest in a number of products that are also dual use (Hamwey 2005).

Dealing with Nontariff Barriers

Tariffs on environmental goods will be easier to tackle than nontariff barriers, which are harder to identify and constantly evolving. Some take the form of "tied aid," in the case of products such as PV systems and wind turbines (Alavi 2007). Exports are associated with aid provided on condition that the recipient country uses the funds to buy goods or services from the donor country, often through a tariff waiver, donations in kind, or directed credit. This practice distorts the condition of competition in favor of the exporter, whose products are granted a tariff preference. Steenblik (2005) points out that the degree of distortion would be less if there were no tariffs to waive in the first place. The distortions caused by tariff waivers for nontied bilateral and multilateral projects are less, especially if system components are purchased through competitive bidding. Steenblik, however, argues that if carried out for too long and too large a scale, tariff waivers would create expectations of further donor grants in the future and drive away capable domestic firms that could develop a strong renewables market on their own.

Often, nontariff barriers are what may be considered measures adopted by countries in the interest of domestic public policy objectives. For example, countries such as Spain and China have put in place local content measures for wind turbines aimed at encouraging domestic production, jobs, and regional development, but these may act as nontariff barriers to foreign imports of wind turbines (Alavi 2007). As noted in chapter 3, other significant nontariff measures that may hamper trade in products such as wind turbines and biofuels include standards and certification requirements, as well as tax and subsidy measures (see Kojima, Mitchell, and Ward 2006).

Dealing with Agricultural Environmental Goods

So far, no WTO member has formally tabled any agricultural product as an environmental good. But Brazil, in its submission (TN/TE/W/59), states that any definition of environmental goods should facilitate a triple-win situation; that is, trade promotion, environmental improvement, and poverty alleviation. Brazil regards improved market access for products with a low environmental impact as contributing to poverty alleviation through income generation and job creation for local populations. It also points out that improved market access for products derived from incorporating cleaner technologies, such as flex-fuel engines and vehicles, could also encourage the use of environmentally efficient products and support the developmental concerns of the developing countries, as these vehicles would use fuels obtained from the processing of natural resources in developing countries.

From a climate change perspective, it would be desirable to reduce barriers to trade in biofuels that contribute (depending on how they are produced) lower GHG emissions compared with fossil fuels (box 4.2). While methanol and biodiesel were proposed by some WTO members as industrial environmental products (biodiesel subsequent to its blending process with a chemical), they were subsequently dropped from a revised list of products submitted by several interested members. If ethanol were included as part of a list for trade liberalization, then agricultural modalities would apply, as opposed to modalities governing industrial products falling under the WTO Negotiating Group on Non-Market Access (NAMA).

The Way Forward on a Possible Agreement on Climate Change Mitigation Products

The success of the ITA had much to do with a critical mass of WTO members being convinced of the economic benefits of liberalizing trade in IT products. A similar critical mass of members who are convinced of the benefits of liberalizing climate-friendly products will need to be created. Despite difficulties plaguing the Doha Round of negotiations, the political timing could not be more appropriate. On this count, it is the stated intent of many—including the EU, the United States, and other countries—to engage in preferential trade agreements. Possibilities also need to be explored for forging regional and/or bilateral agreements on liberalizing climate-friendly technologies.

A number of recent reports by the Intergovernmental Panel on Climate Change have established beyond any doubt the link between human activity and GHG

BOX 4.2

Trade, Environment, and Biofuels

Trade in ethanol and some biodiesel feedstocks is restricted by import and export tariffs and duties in the largest markets. Domestic producers in the European Union, and United States especially, receive additional support through subsidies and duties. Because of these distortions, increasing production volumes in this rapidly growing industry will not be allocated to the most efficient biofuel producers.

Liberalizing trade can produce welfare gains for consumers in industrial countries, where domestic ethanol prices are kept artificially high because of border restrictions—and for efficient producers in developing countries, some of whom could develop a new export industry. Overall, significant efficiency gains could result in a global reallocation of production to the lowest-cost producers. But increasing production of biofuels is also often associated with attendant impacts on food security, deforestation, and biodiversity loss and water use.

Apart from Brazil, it is not clear if other developing countries would benefit from developing biofuel industries. Analysis shows that these circumstances are rare for first-generation technologies and need to be more carefully assessed. High petroleum transport costs could make biofuel production economically viable in some oil-importing countries, even with current technologies, substantially reducing the need for government subsidies. Second-generation technologies, on the other hand, promise a much more favorable balance in terms of environmental and, possibly, social benefits.

Source: Kojima and others 2006.

emissions. The *Stern Review* (2006) has created the necessary momentum globally and has stimulated discussion on the economics of climate change and the need for addressing the issue sooner rather than later. Analysis by IPCC experts, as well as by this study, clearly identify what the necessary goods and technologies for mitigating climate change are, as well as the critical sectors and areas of intervention. This wealth of valuable information and analysis must inform policy makers in deciding how trade and trade negotiations can play a supportive role in mitigation efforts.

From the U.S. side, a recent U.S. Trade Representative report to the Congress suggests that high tariffs and other trade and investment barriers continue to impede access to important GHG-reducing technologies, especially in developing countries (USTR 2006). The report adds that, by reducing the prices of these technologies through substantial reduction or elimination of import tariffs and specific nontariff barriers, developing countries can take concrete and effective action to improve access to products vital for combating pollution, reducing

GHG emissions, and meeting sustainable development goals. The report further adds that the USTR, as a part of the Asia-Pacific Partnership on Clean Development and Climate, is working in a number of countries (Australia, China, India, Japan, and Rep. of Korea) to foster new investment opportunities, build local capacity, and remove barriers to trade in cleaner, more efficient technologies in a variety of settings, including bilateral and regional trade and investment framework agreements, FTAs, and the WTO.

The atmospherics are supported by the fact that EU Trade Commissioner Peter Mandelson emphasized in a recent speech that a WTO-wide deal eliminating all tariffs on trade in green technologies and energy-saving equipment would be the key to finding a business-friendly global solution to climate change. He also indicated that any successor to Kyoto should include the creation of an open global market in environmental technologies and an investment regime supporting green industrial change.

In light of the increased domestic pressure to address climate change more decisively, the EU leaders recently committed to reduce GHG emissions by 2020 by 20 percent below 1990 levels. The high-income OECD countries as a group may not be averse to tackling the issue of liberalizing climate-friendly technologies, both for their self-interest and to bring about faster adoption of these technologies in developing countries.

While still reluctant to bring about broad policy changes relating to climate change, developing countries like China and India—the fastest-growing economies as well as emitters of GHGs—may not actually be averse to getting better access to cleaner technology. The recent Chinese National Climate Change Assessment Report acknowledged that climate change will bring major impacts to several regions of China. However, the report stops short of recommending cuts in China's greenhouse gas output. It says that China should not risk slowing its economic growth by curbing greenhouse gas production. Mexico has already prepared a climate change strategy, and India will prepare a national strategy on climate change before the next round of multilateral negotiations under the UN Framework Convention on Climate Change. The participation of Brazil and other developing countries could hinge on inclusion of biofuels and other agriculture-related products that could provide them with market access.

The Devil Is in the Details: Clarifying Product Coverage, HS Codes, and Product Descriptions

Any possibility for a separate agreement on climate-friendly products hinges on the ability of negotiators to assign clear HS codes or product descriptions for various climate-friendly products and technologies. This has to be done before any agreement is finalized, as the experience with the ITA reveals that any review process would otherwise be hampered.

With regard to climate change technologies, many HS categories at the six-digit level contain both environmental and nonenvironmental goods. Administratively, while it may be easier for WTO members to liberalize the whole HS six-digit category, many members are concerned that this would lead to unintended liberalization of a whole range of products, not just those relevant for environmental purposes. They want liberalization to be confined to six-digit categories that have a single end use.

It is clear that a number of climate-relevant products may be isolated only beyond the HS six-digit level, in which case it will be necessary to harmonize at least the ex-out product descriptions across members. Harmonizing HS codes themselves beyond the six-digit level will be a massive undertaking and would not be viable given the short time horizon for a possible conclusion of the Doha Round, as well as the timing of review cycles of the World Customs Organization (WCO). The WCO considers amendments to the HS once every five years, with implementation taking place from one to two years following notifications to members. The approval of the latest amendment took place in June 2004 and entered into force on January 1, 2007 (Kim 2007).

At least with regard to HS descriptions, it is probably easier to work out harmonization and codes between two countries bilaterally, or between small groups of countries on relevant products. For any change to be multilaterally accepted would need the involvement of all WTO members, a much more complex undertaking. For example, all countries of MERCOSUR apply an eight-digit HS code.

The ITA experience also points to the difficulties created if countries may not have included certain categories within their respective national nomenclatures or are inconsistent in terms of the product category under which they classify ex-outs. These difficulties should be anticipated before any product or product group is specifically considered for inclusion as part of a climate package.

Dealing with Evolving Technology Issues and Nontariff Barriers

Any agreement will also need to include a review mechanism whereby new products would be included after consultation between members, as well as between relevant representatives from the private sector and multilateral environmental agreements—in this case, the UNFCCC Secretariat and international organizations such as the United Nations Environment Programme (UNEP), the World Meteorological Organization (WMO), and the World Bank. In addition, as with all other environmental goods, consultation with the WCO will be important for facilitating long-term implementation of the agreement on climate-friendly products. Similarly, it may not be possible for the agreement to identify and eliminate all NTBs, so a built-in mechanism to enable periodic review and discussion, and to negotiate the elimination of NTBs, will be important.

Prioritizing Products for the Clean Development Mechanism

In the context of the various approaches to liberalization (list, project, or integrated), it may be desirable from a climate change mitigation perspective for WTO negotiators to grant priority for products, technologies, and services imported for projects under the Clean Development Mechanism (CDM). The equipment cost of most renewable energy projects is significantly higher per unit of emission reduction than for other types of potential CDM projects, such as agricultural methane flaring projects (Wilder, Willis, and Curnow 2006). Lowering tariff and nontariff barriers to goods and technologies used in CDM projects could reduce equipment costs and contribute to lower transaction costs for potential investors; of course, those lower costs will need to be complemented by certain measures, such as supportive local regulatory measures.

Further decisions at the 2005 Conference of Parties meeting in Montreal recognized that project activities under various programs could be registered as a single CDM project activity. This implies that a number of initiatives, such as renewable energy projects, would generate a sufficient number of certified emission reductions (CERs) and could be bundled together as part of a programmatic CDM to make transaction costs worthwhile. Trade liberalization of goods used in such programmatic CDMs on renewable energy could complement other measures and incentives to encourage renewable energy projects in developing countries and foster sustainable development.

At the WTO level, goods and services that are an important component of CDM projects should be identified and included in any list of climate-friendly goods and technologies. Alternatively, there could be an understanding or agreement among WTO members that CDM projects, as well as programmatic CDMs, could benefit from automatic approval under a project approach for imports of goods and technologies (as well as services) free from tariffs and nontariff barriers.

Providing Technical and Financial Assistance

Developing countries have perceived many benefits from joining ITA, including employment and a narrowing of the technology gap. But for climate-friendly goods, the primary benefits are global. Consequently, developing countries have fewer incentives to embrace freer trade in climate-friendly goods. To create these incentives, one might call for *smarter* trade as an adjunct to *freer* trade. A simple example: consider bundling tariff reductions on environmental goods with some other benefit to these countries.

Implementation of any agreement on climate-friendly goods and technologies will certainly need to include a package for technical and financial assistance to enable developing countries to deal with implementing liberalization and particularly to deal with challenges created for customs in efficient administration of imports and harmonizing classification. Synergies with regard to technical - assistance within trade facilitation negotiations could also be considered.

In addition to the above recommendations, other technical and financial assistance measures in the context of existing or proposed programs—such as the Integrated Framework or the "Aid for Trade" package, respectively—could also be considered to help countries deal with any adverse shocks of liberalization. Measures could enable them to meet standards and certification requirements and emerge as important and competitive producers and exporters of climate-friendly goods and technologies. A component for trade-related climate change initiatives could be made part of any Aid for Trade package. Assistance from the International Finance Corporation to enable small and medium enterprises to access the latest climate-friendly technologies could also be an important component of a package that is supportive and increases the acceptability of the agreement, particularly among developing countries.

Getting the Right Model

Once the relevant products and technologies for inclusion in any agreement are identified, additional modalities will need to be worked out regarding membership, implementation time periods, and flexible arrangements for developing-country members. (In the ideal scenario, this would be complete elimination of all tariffs within a certain period and eventual elimination of nontariff barriers.) A number of possible models within WTO negotiations could be considered.

The first model could, of course, be liberalization of climate-friendly goods and technologies under the normal course of negotiations on environmental goods through the proposed list, project, or integrated approaches, or some combination of these. While the relevant sectors such as renewable energy or heat and energy management would be highlighted, no separate category of climate-friendly goods would be created.

The second and more innovative approach could be an ITA-type agreement within a single undertaking, whereby members representing a minimum percentage of trade in climate-friendly products would need to join in order for it to come into force. Such an agreement could extend to a subcategory of specifically identified climate-friendly goods within a larger negotiated package of environmental goods or to a stand-alone category (irrespective of whether other environmental goods are liberalized or not). In any case, once the agreement comes into effect, the benefits would extend on an MFN basis to all members—both signatories as well as nonsignatories. The experience of the ITA negotiations reveals that members may be willing to extend benefits on an MFN basis only with a critical mass of members in order to prevent perceived free-riding, particularly by countries that are competitive in the production of goods included in the agreement.

A third option, particularly if negotiations on environmental goods fail to reach a meaningful outcome, would be to consider a plurilateral agreement similar to the Agreement on Government Procurement. In that option, the agreement could come into effect immediately or even independent of the conclusion of the Doha Round negotiations, but only the signatories would extend as well as receive the benefits of trade liberalization in climate-friendly products. The advantage here would be that members, particularly developing countries, need not feel compelled to sign on immediately. It may also provide members that are nonsignatories time to work out harmonized product descriptions or ex-out coding for various products, as well as identify their sensitive products and technical assistance required before they join. Once a critical level of membership is attained for the plurilateral agreement, it could be integrated within the single undertaking, with trade benefits extending on an MFN basis to all members.

Making Tangible Progress Soon in Several Venues

This chapter has outlined some of the key issues and challenges involved in creating a win-win-win opportunity for climate change, trade, and sustainable development through WTO negotiations on environmental goods (figure 4.2). What is important to underline here is that the process need not end with the conclusion of the Doha negotiations. The various challenges and complexities that have been outlined imply the need to deal with a number of issues as part of an ongoing process even beyond the conclusion of the Doha negotiations. This approach could perhaps be done through a built-in mandate for continuous discussions as part of a final Doha agreement. It may even be sold as part of a strategy to save the WTO from the Doha impasse; that is, this is one area where agreement might be negotiated. Such a mandate could address various aspects of the trade–climate change interface that include not only the liberalization of environmental goods and services, but also subsidies and standards, and involve not only the Committee on Trade and Environment but other WTO committees as well.

Just as business as usual in GHG emissions is not sustainable, business as usual in trade negotiations is not an adequate response to the challenges posed in this study. For instance, postponing action until another lengthy round of WTO negotiations following the conclusion of the Doha Round would not be an appropriate response. At least some of the steps mentioned could be taken in the context of the Doha Round, and perhaps even agreed to separately if WTO members fail to come to an agreement and the Doha Round is terminated or suspended indefinitely.

A collapse of the Doha Round could result in a spurt in regional trade agreements (RTAs) as more WTO members seek alternative routes to pursue their trade agendas. What does this imply for trade liberalization in climate-friendly products and technologies? There are opportunities but also other challenges to consider. A number of problems associated with defining environmental and climate-friendly

FIGURE 4.2
Considerations for a Win-Win-Win Package on Trade and Climate Change

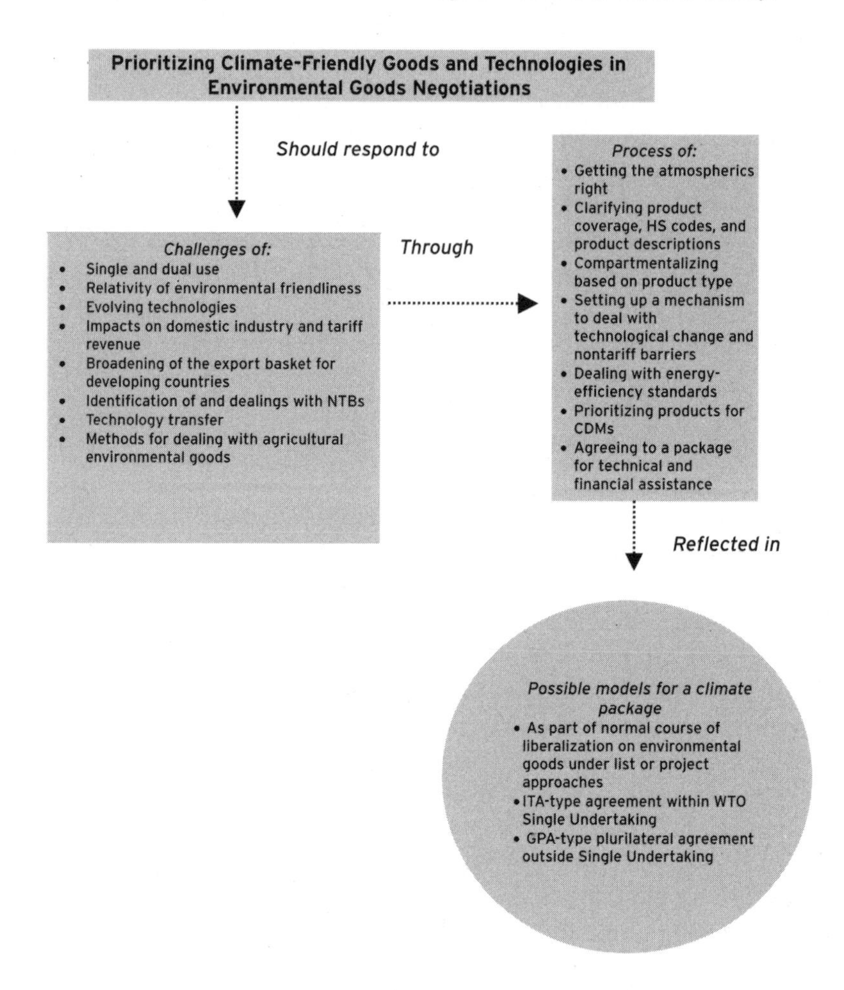

goods, or with determining whether they are dual-use or not, may not arise within RTAs. This is because the objective of most RTAs would be to liberalize the majority, if not all, goods at the HS six-digit level. With regard to provisions aimed at building supply-side capacities, technical assistance, and technology transfer, RTAs may be better suited to include provisions tailored to the needs of participating developing countries. On the other hand, RTAs may also result in diverting trade from countries that are most efficient at producing certain climate-change technologies if these countries are excluded from the RTA.

Although the role of WTO negotiations has been emphasized in this study, there are other venues where progress can be made. In particular, the next

COP/MOP meetings in 2007 and the G-8+5 summit in 2008 both offer opportunities for the leaders of major greenhouse-gas-emitting countries to make specific commitments to reduce tariff and nontariff barriers to international trade and investment in goods, services, and technologies that contribute to the mitigation of climate change.

Key Findings from Chapter 4

- The ongoing WTO negotiations on environmental goods have the potential to contribute significantly to trade liberalization and environmental efforts, but they will need to address a number of challenges.
- The inclusion of specific goods and technologies that are relevant for climate change mitigation may have significant implications with regard to the costs of climate mitigation measures.
- A useful model to draw lessons from could be the Information Technology Agreement previously negotiated in the WTO. An alternative model could be a plurilateral liberalization package for climate-friendly environmental goods on the lines of the Agreement on Government Procurement.
- Political commitment by developed and developing country leaders could contribute to an environment conducive to a meaningful and climate-friendly trade package.

Notes

Acknowledgment. Mahesh Sugathan (ICTSD) contributed to this chapter. The views expressed are those of the author and do not necessarily reflect those of ICTSD as an institution.

1 The language in paragraph 31 (iii) of the Doha Ministerial Declaration is vague and does not define what these goods and services constitute for the purpose of liberalization.

2 Products that did not figure in the WTO submission but could be considered climate friendly and included in any possible future list include solar collector and controller (HS-392510), hydraulic turbines (HS-841011), wind turbine pumps (HS-84138190), small hydel turbines (HS-850239), solar inverter (HS-850440), and compact fluorescent lamps (HS-8539310). Some of these do not conform to six-digit HS level.

3 The "single undertaking" is an important principle of WTO negotiations. It implies that every item of the negotiation is part of a whole and indivisible package and cannot be agreed to separately. In other words, nothing is agreed until everything is agreed. The Agreement on Government Procurement (GPA), together with the Agreement on Trade in Civil Aircraft, constitute the two "plurilateral" agreements in the WTO, which means they extend to only a narrower group of signatories rather than the whole WTO membership.

4 WTO Information Technology Agreement. Introduction and Mandate are available at http://www.wto.org/English/tratop_e/inftec_e/inftec_e.htm.

5 WTO, Note by the Secretariat, G/IT/1/Rev.39, March 26, 2007.

6 WTO Information Technology Agreement. Introduction and Mandate are available at http://www.wto.org/English/tratop_e/inftec_e/inftec_e.htm.

7 WTO, Ministerial Declaration on Trade in Information Technology Products, WT/MIN (96)/16.

8 This is somewhat similar to the ITA, which has been constructed as a dynamic, forward-looking regime explicitly designed to keep up with the rapid pace of technological change in the sector.

CHAPTER 5

Conclusions and Recommendations

Economic growth and poverty reduction require that trading opportunities be rooted in the development agenda of developing countries. As developing countries increasingly fuel global economic growth, countries like India and China—with their increasing share of carbon-intensive development—will be called on to respond to emissions reductions in the post-Kyoto scenario. The Doha negotiations on environmental goods and services provide an opportunity for this growth to leave a smaller carbon footprint than the business-as-usual scenario. In an attempt at advancing the trade and climate change agendas, the key findings of this study are as follows.

Findings

Industrial competitiveness in Kyoto Protocol–implementing countries suffers more from energy efficiency standards than from carbon taxation policies. Though the Kyoto Protocol didn't come into force until 2005, in the 1990s most OECD countries had already established regulatory and fiscal policies, emissions trading systems, and voluntary agreements to combat GHG emissions. Efforts by countries to reduce emissions to meet and exceed Kyoto targets have raised issues of competitiveness in countries that are implementing these policies. The analysis in chapter 2 suggests that efficiency standards are more likely to adversely affect industrial competitiveness than are carbon taxes. Some industries—such as metal products and transport equipment—are more severely affected by the increasing efficiency requirements. For those industries, the analysis also suggests that it does not matter whether such standard requirements are imposed by the exporting country, the importing country, or both.

The effects of carbon taxation policies on industrial competitiveness are often offset by "policy packages." Though competitiveness issues have been much debated in the context of carbon taxation policies, the study finds no evidence that industries' competitiveness is affected by carbon taxes. In fact, the analysis suggests that exports of most energy-intensive industries increase when a carbon tax is imposed by the exporting countries, or by both importing and exporting countries. This finding gives credence to the initial assumption that recycling the taxes back to the energy-intensive industries by means of subsidies and exemptions may be overcompensating for the disadvantage to those industries. A closer examination of specific energy-intensive industries in OECD countries shows that only in the case of the cement industry, has the imposition of a carbon tax by the exporting country adversely affected trade. In the case of the paper industry, trade actually increases as a result of a carbon tax. Results also suggest that trade is not affected when both countries impose the tax.

Some evidence supports relocation (leakage) of carbon-intensive industries to developing countries. A gradual increase in the import-export ratio of energy-intensive industries in developed countries—and a gradual decline in the ratio in some developing regions—indicates that energy-intensive production is gradually shifting to developing countries as a result of many different factors, including climate change measures in developed countries. Although the trend is converging, the import-export ratio is still greater than 1 in developing countries and less than 1 for developed countries, suggesting that developing countries continue to be net importers of energy-intensive products. Lack of strong evidence of relocation suggests that while the overarching objective of climate policies is to reduce emissions, these policies have been designed to shield the competitive sectors of industrialized economies. More stringent climate policies in industrialized countries in the future may continue to provide the necessary impetus for a more visible leakage of carbon-intensive industries. .

Trade measures can be justified only under certain conditions. If a country adopts a border tax measure or even resorts to an outright import ban on products from countries that do not have carbon restrictions, such measures could be in violation of the WTO rules unless they can be justified under the relevant GATT rules. Articles XX(b) and (g) allow WTO members to justify GATT-inconsistent measures, either if these are necessary to protect human, animal, or plant life or health, or if the measures relate to the conservation of exhaustible natural resources, respectively. However, Article XX requires that these measures not arbitrarily or unjustifiably discriminate between countries where the same conditions prevail, nor constitute a disguised barrier to trade. Since most climate change measures do not directly target any particular products, but rather focus on the method by which greenhouse gases may be implicated related to production, issues related to process and production methods (PPMs) are critical for the compatibility between the WTO and Kyoto regimes. In the recent Shrimp-Turtle dispute, the WTO Dispute Settlement Panel and the Appellate Body may have opened the doors to the permissibility of trade measures based on PPMs.

The proposed EU "Kyoto tariff" may hurt the United States' trade balance. There is increasing industry pressure in the EU to sanction U.S. exports for not adhering to the Kyoto targets. This has resulted in calls for a Kyoto tariff on a range of U.S. products to compensate for the loss in competitiveness. Simulation analysis undertaken for this study finds that the potential impact of such punitive measures by the EU could result in a loss of about 7 percent in U.S. exports to the EU. The energy-intensive industries such as steel and cement, which are the most likely to be subject to these provisions and thus would be most affected, could suffer up to a 30 percent loss. Actually, these are conservative estimates, given that they do not account for trade diversion effects that could result from the EU shifting to other trading partners whose tariffs could become much lower than the tariffs on the United States.

Varied levels of tariff and nontariff barriers (NTBs) are impediments to the diffusion of clean energy technologies in developing countries. While the current Kyoto commitments for GHG emissions reduction apply only to Annex I countries, the rising share of developing-country emissions resulting from fossil fuel combustion will require future commitment and participation of developing countries, particularly large emitters like China and India. Some developing countries have already taken measures to unilaterally mitigate climate change; for instance, they have increased expenditures on R&D for energy efficiency and renewable energy programs. It is important that these countries identify cost-effective policies and mitigation technologies that contribute to long-term low-carbon growth paths. Especially for coal-driven economies like China and India, investments are critical in clean coal technology and renewable energy such as solar and wind power generation. Detailed analysis undertaken for the study in chapter 3 suggests that varied levels of tariffs and NTBs are a huge impediment to the transfer of these technologies to developing countries. For example, energy-efficient lighting in India is subject to a tariff of 30 percent and a nontariff barrier equivalent of 106 percent.

Recommendations

A closer examination of the "policy bundle" or package associated with energy taxation is warranted. The results emerging from the analysis in chapter 2 suggest that carbon taxation policies do not adversely affect the competitiveness of energy-intensive industries. This finding suggests that complementary policies (implicit subsidies, exemptions, etc.)—which are used in conjunction with carbon taxation policies levied by Kyoto Protocol-implementing countries, particularly on energy-intensive industries—could be negating any impact of carbon taxation. A more detailed study of this issue is warranted, as it will yield a greater understanding of the implicit subsidies or costs that are associated with each industry. The importance of this finding cannot be understated, as trade measures are justified based on perceptions of higher costs for energy-intensive industries in developed countries and associated loss of competitiveness on account of these costs. The political economy of carbon taxation policies may be used to gain greater insights into the policy package as well.

It would be useful at the outset for trade and climate regimes to focus on a few areas where short-term synergies could be exploited. The energy efficiency and renewable energy technologies needed to meet future energy demand and reduce GHG emissions below current levels are largely available. The WTO parties can do their part by seriously considering liberalizing trade in climate-friendly and energy-efficient goods as a part of the ongoing Doha negotiations to support Kyoto. Within the UNFCCC, it would also help to accelerate and bring greater clarity to the technology transfer agenda. Within the Kyoto Protocol, the most important priority regarding the linkage to trade would be to facilitate a uniform approach to the pricing of greenhouse gas emissions.

Removal of tariff and nontariff barriers can increase the diffusion of clean technologies in developing countries. As stated above, access to climate-friendly clean energy technologies is especially important for the fast-growing developing economies. Within the context of current global trade regime, the study finds that a removal of tariffs and NTBs for four basic clean energy technologies (wind, solar, clean coal, and efficient lighting) in 18 of the high-GHG-emitting developing countries will result in trade gains of up to 13 percent. If translated into emissions reductions, these gains suggest that—even within a small subset of clean energy technologies and for a select group of countries—the impact of trade liberalization could be reasonably substantial.

Streamlining of intellectual property rights, investment rules, and other domestic policies will aid in widespread dissemination of clean technologies in developing countries. Firms sometimes avoid tariffs by undertaking foreign direct investment (FDI) either through a foreign establishment or through projects involving joint ventures with local partners. While FDI is the most important means of transferring technology, weak intellectual property rights (IPR) (or perceived weak IPR) regimes in developing countries often inhibit diffusion of specific technologies beyond the project level. Developed country firms, which are subject domestically to much stronger IPRs, often transfer little knowledge along with the product, thus impeding widespread dissemination of the much-needed technologies. Further, FDI is also subject to a host of local country investment regulations and restrictions. Most non–Annex I countries also have low environmental standards, low pollution charges, and weak environmental regulatory policies. These are other hindrances to acquisition of sophisticated clean energy technologies.

The huge potential for trade between developing countries (South-South trade) in promoting clean energy technology in those countries needs to be explored more. Traditionally, developing countries have been importers of clean technologies, while developed countries have been exporters of clean technologies. However, as a result of their improving investment climate and huge consumer base, developing countries are increasingly becoming major players in the manufacture of clean technologies. A key development in the global wind power market is the emergence of China as a significant player, both in manufacturing and in investing in additional wind power capacity. Similarly, other developing countries have emerged as manufacturers of renewable energy technologies. India's photovoltaic (PV) capacity has increased several times in the last four years, while Brazil continues to be a world leader in the production of biofuels. These developments augur well for a buoyant South-South technology transfer in the future.

Clean technology trade would greatly benefit from a systematic alignment of harmonization standards. The volume of trade and the level of tariffs can be examined by identifying and tracking the unique HS code associated with each technology or product under the Harmonized Commodity Description and Coding System (commonly called the harmonized system or HS). Typically, each component of the technology has a different HS code. At the WTO-recognized six-digit code level, clean energy technologies and components are often found lumped together with other technologies that may not necessarily be classified as being beneficial to either the global or even local environment. Solar photovoltaic panels are categorized as "Other" under the subclassification for light-emitting diodes (LEDs). Such categorization suggests that reducing the customs tariff on solar panels might also result in tariff reduction for unrelated LEDs. Similarly, clean coal technologies and components are not classified under a separate category, and all gasification technologies are lumped together. The imprecise definition also raises another issue for countries that are considering removal of trade barriers to clean energy equipment and components. In cases where the codes are not detailed enough, the scope of the tariff reduction may become much broader than anticipated.

The ongoing WTO negotiations on environmental goods have the potential to contribute significantly to both trade and climate change efforts, but the negotiations will need to address a number of challenges. Liberalizing trade in specific goods and technologies that are relevant for climate change mitigation may have implications with regard to the costs of mitigation measures, particularly those technologies that face high tariff and nontariff barriers to trade. The relevant concerns cannot be disregarded, such as those related to definition of relevant products (especially products that also have nonenvironmental uses); harmonizing classifications and descriptions across countries within the harmonized system; changes in technology; issues related to perceived impacts on domestic industries; and nontariff measures and access to technology. Goods that would benefit include those that directly address climate change mitigation, as well as environmentally preferable products that contribute to zero or reduced GHG emissions during production, consumption, or use. Goods and technologies used in CDM projects (including programmatic CDMs) are particularly relevant.

Political economy dynamics may necessitate the consideration of innovative packages for trade liberalization in climate-friendly goods. One could be an ITA-type agreement within a single undertaking, whereby members representing a minimum percentage of trade in climate-friendly products would join. Such an agreement could be a subcategory within any larger negotiated package of environmental goods or in a separate agreement. A second option, particularly if negotiations on environmental goods fail to reach a meaningful outcome, would be to consider a plurilateral agreement similar to the agreement on government procurement. In that option, the agreement could come into effect immediately or even independent of the conclusions of the Doha Round negotiations, but only the signatories would extend as well as receive the benefits of trade liberalization in climate-friendly products. The advantage here would be that members, particularly developing countries, need not feel compelled to sign on immediately.

RTAs also offer opportunities but there are challenges to consider. A collapse of the Doha Round could result in a spurt in regional trade agreements (RTAs) as more WTO members seek alternative routes to pursue their trade agenda. A number of problems associated with defining environmental and climate-friendly goods will be less of an issue as most RTAs would normally liberalize at a broader HS level (usually six-digit). With regard to provisions aimed at building supply-side capacities and technical assistance, RTAs may be better suited to include provisions tailored to the needs of participating developing countries. On the other hand, RTAs may also result in the diversion of trade from countries that are most effective at producing climate-friendly technologies if those countries are excluded from an RTA.

Making tangible and immediate progress is necessary in several venues. Just as business as usual in GHG emissions is not sustainable, business as usual in trade negotiations is not an adequate response to challenges posed in the study. At least some of the steps mentioned could be taken in the context of the Doha Round, and perhaps even agreed to separately if WTO members fail to come to an agreement and the Doha Round is terminated or suspended indefinitely. Although the role of the WTO negotiations has been emphasized in this study, there are other venues where progress can be made. In particular, the next COP/MOP meetings (Conference/Meeting of the Parties to the Protocol) in 2007 and the G-8+5 summit in 2008 both offer opportunities for the leaders of the major GHG-emitting countries to make specific commitments to reduce tariff and nontariff barriers to international trade and investment in goods, services, and technologies that contribute to the mitigation of climate change.

APPENDIX 1

Kyoto Protocol: Countries Included in Annex B to the Kyoto Protocol and Their Emissions Targets

Country	Target (1990**/ 2008/2012)
EU-15*, Bulgaria, Czech Republic, Estonia, Latvia, Liechtenstein, Lithuania, Monaco, Romania, Slovakia, Slovenia, Switzerland	–8%
United States***	–7%
Canada, Hungary, Japan, Poland	–6%
Croatia	–5%
New Zealand, Russian Federation, Ukraine	0
Norway	+1%
Australia	+8%
Iceland	+10%

* The EU's 15 member states will redistribute their targets among themselves, taking advantage of a scheme under the Protocol known as a "bubble." The EU has already reached agreement on how its targets will be redistributed.

** Some EITs have a baseline other than 1990.

*** The United States has indicated its intention not to ratify the Kyoto Protocol.

APPENDIX 2

Measures to Combat Climate Change

Regulatory Measures

Regulatory instruments, such as regulations, standards, directives, and mandates, have been most commonly used to promote energy efficiency and renewable energy, including cogeneration and low-emission motor vehicles in OECD countries. Some of the more prominent initiatives in place include the following:

- The EU Renewable Electricity Directive of 2001 (Directive 2001/77/EC), which set a target to increase the share of renewable energy production (such as wind, solar, geothermal, wave, tidal, hydroelectric, biomass, landfill gas, sewage treatment gas, and biogas energies) to 12 percent of total energy use, and of renewable electricity production to 22 percent of total electricity consumption in 2010, with specific targets for each member state.[1] In March 2007, European leaders revised this and agreed to have 20 percent of their overall energy needs covered by renewables. In order to give member states more flexibility, the Commission did not put forward specific subtargets, such as for renewable electricity or heating and cooling.
- The *Renewables Obligation* enacted in the United Kingdom requires suppliers to source a specific and annually increasing percentage of electricity they supply from renewable sources, to meet a target of 10 percent of electricity from renewable sources by 2010.[2] Other countries, including Austria, Belgium (the regions

of Flanders and Wallonia), Italy, Netherlands, and Sweden, have also adopted minimum renewable energy targets, and some combined them with tradable renewable energy certificates (TRCs) as the United Kingdom did.

- In the EU, a directive on combined heat and power (CHP; Directive 2004/8/EC) was agreed to in 2004 that provides a framework for promoting and developing high-efficiency cogeneration.[3] In 2000, the U.K. government set a new target to achieve at least 10,000 MWe of installed good-quality CHP capacity by 2010.
- The Environmental Code in Sweden (1999) stipulates that the best possible technology should be used in all industrial operations. It also states that anyone running an operation or implementing a measure should conserve raw materials and energy, and that recovery and recycling should be conducted when possible.
- In Japan, the revised Energy Conservation Law, in force since April 1999, sets energy conservation standards for home/office appliances and fuel efficiency for automotives on the basis of the most efficient products available on the market, in contrast to the generally accepted approach to set these standards on the basis of the average efficiency within the product class.[4] The Top Runner program has been effective in stimulating the diffusion of existing efficient technologies and enhancing the industrial competitiveness of Japanese products (UNFCCC 2005).
- The EU's Energy Performance of Buildings Directive (Directive 2002/91/EC) requires member states to adopt energy performance standards and introduces energy labeling of buildings across the EU, along with a requirement to evaluate the opportunities for installing renewable energy systems in buildings above a certain size.
- Under the EU's directive on energy labeling of domestic household appliances (Directive 96/75/EC), domestic household appliances sold in the EU must carry a label grading them according to their energy efficiency, with grades running from A (high energy efficiency) to G (low efficiency) to allow consumers to choose the most efficient ones. New Zealand will implement a similar regulation by 2008 that imposes a requirement to display energy efficiency labels to ensure that certain types of products meet minimum standards of energy efficiency.
- Canada has just recently (April 26, 2007) announced an aggressive strategy to tackle climate change. *Turning the Corner: A Plan to Reduce Greenhouse Gases and Air Pollution* aims to cut greenhouse gas emissions per unit of production by 18 percent by 2010. This plan sets mandatory reduction targets for major industries that produce greenhouse gases, but it allows companies to choose the method by which to meet their reduction targets. Methods include reducing emissions in their facilities, investing in emissions-reducing technologies, or participating in domestic emissions trading schemes and the Kyoto Protocol's Clean Development Mechanism.

Fiscal Measures

Many OECD countries have explicitly adopted a range of fiscal policies and measures, including environmental taxes and subsidies, as part of policy packages developed to implement the Kyoto commitments. All OECD countries have introduced some kind of environmental taxation, and an increasing number of countries are implementing comprehensive green-tax reforms.[5]

Seen by many as one of the most cost-effective instruments for environmental objectives, carbon/energy taxes (taxes based on the carbon or energy content of the energy products) are among the most widely used environmental tax instruments, especially in Northern Europe. A number of OECD countries—including Denmark, Finland, Germany, the Netherlands, Norway, Slovenia, Sweden, and the United Kingdom—employ carbon or energy taxes. Tax rates vary markedly across the countries, thus the average price of a ton of carbon is somewhat different from country to country. These taxes usually vary both across different fossil fuel categories (e.g., fuel oil, natural gas, electricity, liquefied petroleum gas) and across sectors (e.g., household, industrial) and sometimes also by size of use and geographical location.

All countries that have introduced carbon/energy taxes have also introduced special tax reductions, rebates, tax ceilings, or exemptions in order to address concerns about the effect of the taxes on industrial competitiveness (especially in energy-intensive industries), which in turn reduces the economic impact and environmental effectiveness of the instrument.[6] Some of the more prominent fiscal initiatives are listed below:

- In 1990, Finland was the first country to introduce a carbon tax initially with few exemptions for specific fuels and sectors. The tax was based on the CO_2 content of the fuel, starting at a comparatively low level of Mk 6.7 per ton of CO_2 (US$1.2/t CO_2). Since then, however, the tax has been changed many times, from a low but "pure" carbon tax to a much higher but much less CO_2-related tax, and further exemptions have been added (OECD 1997).
- In 1991, Sweden introduced a carbon tax and a value added tax on energy, and lowered the existing energy tax, as part of an overall fiscal reform. The original tax rates varied according to the average carbon content of different fossil fuel types, but they were applied equally across basic uses (household and nonmanufacturing industry) and industries, placing a tax of SEK 0.25/kg (US$100/t CO_2) on oil, coal, natural gas, LPG, gasoline, and fuel for domestic air transportation. In 1993, however, the industry rate was reduced to one-quarter of the new basic rate in order not to hamper international competitiveness of the industry sector. Further reductions for energy-intensive enterprises were also taken; for example, until 2004, industrial consumers paid no energy tax and only 50 percent of the general carbon tax (OECD 1997; Fouquet and Johansson 2005).

- The Norwegian authorities introduced carbon taxes in 1991. The tax rate differed across fossil fuel categories and the geographic location of the activity (mainland and offshore). In 1996, the tax per ton of CO_2 ranged from US$17 on petroleum coke to $55.60 on gasoline and on gas use in the North Sea (OECD 1997). The present carbon tax scheme is mainly based on the sale of fossil fuel products. Process emissions from several export-oriented mainland manufacturing industries, including aluminum and chemicals, have been exempted. Exemptions are also granted to the fishing fleet, aviation, coastal shipping of goods, and international shipping. As a result, only about 60 percent of the CO_2 emissions and only about 20 percent of emissions from manufacturing are subject to the tax (OECD 1999).
- The EU carbon/energy tax proposal was one of the EU's early policy responses to its signing of the UNFCCC.[7] However, the proposal proved contentious, was amended in 1995, and was eventually withdrawn by the Commission in 2001 (European Energy Agency 2004). After years of negotiations, an agreement was reached in the EU on a minimum tax directive concerning energy products and electricity (Directive 2003/96/EC), and the directive entered into force in the beginning of 2004. The directive extends the EU's minimum rates of taxation, previously confined to mineral oils, to all energy products, including coals, natural gas, and electricity. Although many EU member states have already set higher national taxes on energy products than the ones set by the EC in the directive, some member states are required to introduce or increase energy taxes. Special tax provisions are provided if companies participate in either a voluntary agreement or a tradable permit scheme. In addition, the commercial use of energy products is subject to lower tax rates.

Some countries use part or all of the tax revenues to offset the negative effects of the taxes to reduce distortions in labor or capital markets and to address international competitiveness. In the United Kingdom, most of the proceeds of the Climate Change Levy are allocated to reducing distortionary labor taxes, such as employers' national insurance, in the form of employment tax refunds. The tax revenues are also being recycled to additional government support for energy efficiency measures via the Carbon Trust.[8] The revenues from the German ecotax are almost fully returned to the taxpayers by using them for a graduated reduction of employer-employee pension contributions. Danish carbon tax revenues from industry are entirely recycled in that sector through lower employers' social security contributions, investment grants for energy efficiency improvements, and a fund for small businesses.

Various forms of other fiscal instruments, including subsidies, tax credits, and feed-in tariffs, have been widely used to support and encourage energy efficiency, renewable energy sources, and low-carbon technologies.[9] Canada, Italy, Japan, and Sweden have adopted this type of measure, mostly targeting the energy and electricity generation sectors and the building/residential and transport sectors. In the building sector, grants and subsidies are usually focused on promoting renewable energy systems for space and water heating (e.g., subsidies for biomass and biogas district heating in Austria). For energy production, feed-in tariffs for renewable energy sources were introduced in France, Germany, Ireland, the Netherlands, Spain, and Switzerland. For example, Ireland's Renewable Energy Feed-In Tariff (REFIT) program guarantees power prices for all registered renewable power generators to attract "sufficient confidence for investment finance and loan capital which may not otherwise be provided." Tariffs range from 5.7 eurocents/kWh for large wind-farm power to 7.2 eurocents/kWh for biomass energy.[10]

Market-Based Instruments

Market-based instruments, especially emissions trading and tradable renewable energy certificates (TRCs), are becoming increasingly important climate change strategies as effective means to help decrease the cost of mitigating GHG emissions (IEA 2001). Emissions trading has been used since the 1980s to control non-greenhouse-gas emissions. Recently, it has been used to address greenhouse gas emissions, including CO_2. Several countries have been implementing or discussing domestic emission trading systems. Trading systems for domestic greenhouse gas emissions are implemented in Denmark, the United Kingdom, Norway, France, Japan, and the EU, and several countries, including Switzerland, the Slovak Republic, and Canada, are considering implementing them.[11] Each of these systems has different designs, covers different sectors, and has different methods of allocation. Some of the more prominent initiatives include the following:

- The U.K. Emissions Trading Scheme (ETS) is the first economy-wide GHG emissions trading scheme. The scheme was launched in March 2002 and runs until December 2006, with final reconciliation in March 2007. Thirty-three organizations ("direct participants") have voluntarily taken on emissions reduction targets to reduce their emissions by 3.96 million tons of CO_2-equivalent (t CO_2-e) by the end of the scheme. Over the first three years (2002–04), the U.K.'s ETS delivered emissions reductions of 5.9 million t CO_2-e. The U.K. ETS also helped to shape the design and implementation of the EU ETS under the oversight of Department for Environment, Food and Rural Affairs.[12]
- The EU ETS is the largest company-level trading system for CO_2 emissions in terms of its value and volume, and one of the major policy tools to reduce emissions in the EU. The first trading period is from 2005 to 2007, and the second from 2008 to 2012. The scheme covers mainly energy-intensive industries (e.g., power and heat generators, oil refineries, ferrous metals, cement, and pulp and paper), 12,000 installations in six sectors of the EU-25, representing

about half of the CO_2 emissions from the EU-25 (Commission of the European Communities 2005). The scheme is a cap-and-trade system; the installations are allocated permits by governments that allow them to emit a certain amount of CO_2 each year. Those that emit less than their allocation can sell the surplus allowances, and those that expect to emit more than their allowance have the option of either investing in ways to reduce their emissions or buying additional allowances on the market. Companies can also use credits from Kyoto's project-based mechanisms (Joint Implementation and CDM) to fulfill their obligations under the scheme. The EU ETS was worth US$8.2 billion in 2005, which corresponded to 322 million t CO_2-e, and traded $6.6 billion in just the first three months of 2006 (World Bank and IETA 2006).

- Launched in 2005, the Japanese voluntary emission trading scheme seeks measures to achieve certain and cost-efficient emissions reductions and to accumulate knowledge and experience in domestic ETS. The government selects target facilities from applicants based on the cost-effectiveness of greenhouse gas emissions reduction activities. Thirty-four companies and corporate groups were selected as participants in the scheme. Subsidies are provided for installation of new facilities for the reduction of emissions in return for pledging a certain amount of GHG emissions reduction, but if the participants fail to achieve the targets, they have to return the subsides to the government.[13] The total of emissions reductions promised by the facilities for fiscal 2006 is 276,380 tons, or 21 percent of their average annual CO_2 emissions in the base years, fiscal 2002 to 2004 (IETA 2005).

Domestic emissions trading schemes are often used in policy packages with taxes and voluntary agreements. For instance, the U.K. ETS is open to the companies who signed on to the Climate Change Agreement. These companies can use the scheme either to buy allowances to meet their emissions targets, or to sell any overachievement of these targets.

The tradable renewable energy certificate (TRC) system, also called green tags or renewable energy credits, supports the production of renewable energy. The TRC system obliges energy producers to supply customers with a percentage of renewable energy (green quotas) and then allows the quotas or certificates to be traded (independent of the physical energy) on special certificate markets (IEA 2000). The U.K. Renewable Obligation, for instance, is to be facilitated by allowing the trade of renewable obligation certificates so that electricity sellers overcomplying with the target can sell the certificates to those who undercomply.

A number of subsovereign and subnational initiatives also exist, including those enacted by the state of California,[14] Regional Greenhouse Gas Initiative (RGGI) of the Northeastern States,[15] and New South Wales Greenhouse Plan. Although these are done at the subsovereign and subnational levels, they may have important implications for international trade.

Voluntary Agreements (VAs)[16]

The insufficiency of other policies and measures to achieve meaningful reductions and meet national emissions commitments had led to the search for more innovative solutions, particularly through engaging the private sector in the mitigation process.[17] The number of VAs related to energy efficiency or GHG emissions reductions increased sharply in OECD countries since the UNFCCC. Energy and manufacturing sectors led other sectors of the economy with such measures. VAs differ from other measures in that they are negotiated directly between government and industry or firms rather than resulting from mandates imposed by the governments. They are often the preferred policy approach from industry's perspective since they leave more of the initiative with industry and offer more flexibility. The consequences for noncompliance vary considerably among agreements. Some VAs have strict binding targets (e.g., the U.K.'s Climate Change Levy), while others have no penalties for failure to attain the stated target (e.g., Finland's Agreements on the Promotion of Energy Conservation in Industry).

VAs are often used in policy packages with other policy instruments, such as regulations, taxes, and tradable permit schemes. Governments frequently provide incentives to draw out participation by industry in VAs. For example, the U.K. Climate Change Levy scheme includes climate change agreements with energy-intensive sectors, which provides for an 80 percent discount of the levy if commitments are being made to improve energy efficiency and to reduce environmental impact. Similar provisions can be found in the Danish carbon taxation system. In Denmark, companies with energy-intensive processes get a tax reduction if they enter an individual agreement with the Danish Energy Agency. By 2001, more than 300 firms, accounting for about 60 percent of total energy consumption by industry, had concluded an agreement (IEA 2002). In Switzerland, priority is given to voluntary action in lowering fossil fuel consumption (the CO_2 Act in 1999), but if voluntary and other measures are not sufficient, the Federal Council is authorized to impose an incentive CO_2 tax. As soon as the CO_2 tax is introduced, VAs will be transformed into legally binding commitments, and companies not complying with their reduction target will be penalized (Swiss Confederation 2005).

In some countries, VAs are the main climate change measures and are expected to be highly effective in achieving energy and greenhouse gas reductions in industries. Some of the more prominent initiatives are listed below:

- In Japan, most of the initiatives related to CO_2 reduction are voluntary, since VAs are preferred because they offer lower institutional obstacles.[18] Japan's voluntary action plan, Wisdom of Industry, covers 82 percent of CO_2 emissions from the industry/energy conversion sectors (34 subsectors) and is expected to deliver about 30 percent of the needed energy savings and the related emission savings. Ensuring the success of VAs requires continuous efforts to promote public awareness. The success of the plan stems in part from the government's involvement. The Japanese government reviews progress periodically, and the process of reviewing efforts is quite transparent to the public (UNFCCC 2005).

- In the Netherlands, VAs, in combination with fiscal incentives and environmental permits, are the main policy tool used to limit industry GHG emissions. Companies that account for almost all (96 percent) of Dutch industrial energy use have agreed with an energy efficiency "benchmarking covenant." Under the covenant, these companies pledge to be among the world leaders in energy performance and thus contribute to the effective implementation of the Kyoto Protocol. The German Third National Communication to the UNFCCC also indicates that VAs in industry are expected to have a greater GHG impact than any other policy instruments in reducing GHG emissions by 2010 (OECD 2003).
- The voluntary commitments by European, Japanese, and Korean carmakers to reduce CO_2 emissions from cars sold in the EU by 25 percent in 2008/09 in relation to 1995 is the first pillar of the EU strategy to reduce CO_2 emissions from passenger cars. The scheme is expected to have significant effects.

Notes

1. Common and Coordinated Policies and Measures (CCMPs) are a central part of the EU's climate strategy. At the European level, a comprehensive package of policy measures to reduce greenhouse gas emissions have been initiated through the European Climate Change Programme (ECCP). Each of the EU member states has also implemented its own domestic actions that complement the ECCP measures.
2. The level of the obligation in England, Wales, and Scotland is 4.9 percent for 2005–06, rising to 15.4 percent by 2015–16.
3. Combined heat and power (CHP), also known as cogeneration, is a very efficient technology for generating electricity and heat together that, unlike conventional forms of power generation, puts to use the by-product heat that normally leaves the environment.
4. According to the law, standards are voluntary for manufacturers and retailers, but no manufacturer would risk negative publicity because it failed to achieve the standards set.
5. OECD (2001). In Denmark, Italy, the Netherlands, and Sweden, carbon/energy taxes were introduced as part of the reform of existing energy and other taxes to take account of environmental considerations.
6. The fear of reduced international competitiveness in energy-intensive sectors is one of the major obstacles to the implementation of environmental taxes. In 1992, the European Commission presented a proposal for a carbon/energy tax, which included the exemption of the six most energy-intensive industrial sectors. But the proposal was abandoned in 2001, in part because of strong business opposition.
7. The European Commission has a long-term objective to further harmonize minimum levels of tax rates across the EU.

8 The Carbon Trust was launched in 2001 as a component of the Climate Change Levy package. The aims of the trust are to encourage the research and development of low-carbon technologies and energy-saving measures.

9 Feed-in tariffs set a predetermined buy-back rate for an amount of electricity produced.

10 The Irish government will fund the feed-in tariff in compliance with the EU Directive on Electricity Production from Renewable Sources, intending to generate 13.2 percent of its energy from renewable sources by 2010. http://www.iea.org/textbase/pamsdb.

11 There have been serious negotiations recently to link the EU ETS with a Californian GHG trading scheme. There have also been plans to launch state-level trading schemes in other U.S. states.

12 More details are available at: http://ec.europa.eu/environment/climat/emission/ mrg_en.htm.

13 About one-third of the cost of the emissions reduction activities will be subsidized by the government as incentive. The total government budget for the subsidy is 2,596,340,000 yen (about US$23.6 million).

14 California recently passed a bill that requires the state's major industries—such as utility plants, oil and gas refineries, and cement kilns—to reduce their emissions of carbon dioxide and other greenhouse gases by an estimated 25 percent by 2020. California's emission regulations for passenger vehicles were already above the federal limits.

15 The Regional Greenhouse Gas Initiative, or RGGI, is a cooperative effort by Northeastern and Mid-Atlantic states to reduce CO_2 emissions. The RGGI participating states will be developing a regional strategy for controlling emissions. This strategy will more effectively control greenhouse gases, which are not bound by state or national borders. Central to this initiative is the implementation of a multistate cap-and-trade program with a market-based emissions trading system. The proposed program will require electric power generators in participating states to reduce CO_2 emissions.

16 According to the Intergovernmental Panel on Climate Change (IPCC 2001), a voluntary agreement is "an agreement between a government authority and industry to achieve environmental objectives or to improve environmental performance beyond compliance."

17 Voluntary agreements are popular due to their lower cost, flexibility, and greater political consensus compared with regulatory and fiscal instruments. They reflect the increasing reluctance of governments to impose regulatory or fiscal policies on firms that must compete internationally (OECD 2005; IEA 2000).

18 The Japanese government, as well as the public, successfully put pressure on the *kendanren*, a Japanese business association that coordinates these voluntary initiatives (IEA 2002).

APPENDIX 3

Model Specification and Results

TO STUDY THE EFFECTS OF ENVIRONMENTAL MEASURES on export performance, a standard gravity equation (Feenstra 2003) is used. The log of industry-level bilateral exports between two countries is regressed relative to the product of the two GDPs, on importer fixed effects (α_i), exporter fixed effects (α_j), year fixed effects (α_t), product fixed effects (α_k), the log of distance between the two countries (dist), dummy variables on common borders (border), common currency, and common free trade agreements (FTAs).

Based on the year a carbon tax is implemented in a country, three dummy variables (ct1, ct2, and ct3) are constructed. The first one is if only an exporting country has a carbon tax in the year; the second is if only an importing country has a carbon tax in the year; and the third is if both countries have a carbon tax in the year. The coefficients of these carbon tax dummy variables capture the change in exports relative to the baseline scenario when neither importing nor exporting countries has a carbon tax. Similarly, based on the year an energy efficiency standard is implemented in a country, three dummy variables (ees1, ees2, and ees3) are constructed to capture the effects on exports relative to the baseline scenario when no such standard is in place.

The basic model therefore is:

$$\ln\left(\frac{\text{export}_t^{kij}}{\text{GDP}_t^i + \text{GDP}_t^j}\right) = \alpha_i + \alpha_j + \alpha_t + \alpha_k + \beta_1 \ln \text{dist}^{ij} + \beta_2 \text{border}^{ij}$$
$$+ \beta_3 \text{currency}_t^{ij} + \beta_4 \text{FTA}_t^{ij} + \gamma_1 \text{ct1}_t^i + \gamma_2 \text{ct2}_t^j + \gamma_3 \text{ct3}_t^{ij}$$
$$+ \delta_1 \text{ees1}_t^i + \delta_2 \text{ees2}_t^j + \delta_3 \text{ees3}_t^{ij}$$

The results are presented in tables 3A, 3B, and 3C.

TABLE 3A

Results from the Competitiveness Analysis: Effects of Climate Change Measures on all Relevant Industries

Dependent variable: Log of bilateral export relative to the product of GDP in two countries

	(1)	(2)	(3)	(4)
Log of bilateral distance (km)	−1.387*** (0.018)	−1.386*** (0.018)	−1.387*** (0.018)	−1.387*** (0.018)
Common border dummy variable	0.961*** (0.051)	0.963*** (0.050)	0.963*** (0.051)	0.963*** (0.051)
Common currency dummy variable	0.171*** (0.036)	0.173*** (0.037)	0.174*** (0.036)	0.174*** (0.036)
FTA dummy variable	0.408*** (0.069)	0.412*** (0.069)	0.409*** (0.069)	0.409*** (0.069)
ct1	0.034 (0.033)		0.029 (0.033)	−0.051 (0.034)
ct2	−0.040* (0.024)		−0.043* (0.024)	−0.016 (0.023)
ct3	−0.013 (0.045)		−0.017 (0.045)	−0.071 (0.048)
ees1		−0.105*** (0.036)	−0.102*** (0.036)	−0.075** (0.038)
ees2		−0.090*** (0.033)	−0.093*** (0.033)	−0.062* (0.035)
ees3		−0.099*** (0.036)	−0.100*** (0.036)	−0.027 (0.037)
ct1*energy-intensive input industry				0.462*** (0.022)
ct2*energy-intensive input industry				−0.151*** (0.034)
ct3*energy-intensive input industry				0.317*** (0.036)
ees1*energy-intensive output industry				−0.154*** (0.044)
ees2*energy-intensive output industry				−0.172*** (0.049)
ees3*energy-intensive output industry				−0.402*** (0.041)

TABLE 3A

Results from the Competitiveness Analysis: Effects of Climate Change Measures on all Relevant Industries *(continued)*

	(1)	(2)	(3)	(4)
Constant	−28.044*** (0.217)	−27.963*** (0.217)	−27.961*** (0.216)	−28.007*** (0.215)
Exporting country fixed effects	Yes	Yes	Yes	Yes
Importing country fixed effects	Yes	Yes	Yes	Yes
Industry fixed effects	Yes	Yes	Yes	Yes
Year fixed effects	Yes	Yes	Yes	Yes
Sample size	307,957	307,957	307,957	307,957
R-squares	0.6103	0.6103	0.6104	0.6114

Note: *, **, *** indicate statistical significant at 10%, 5%, and 1% level, respectively.

Standard errors in parentheses are clustered by country-year pair.

Sample is pooled across all three-digit ISIC manufacturing industries.

TABLE 3B

Results from the Competitiveness Analysis: Effects of Climate Change Measures on Energy-Intensive Industries

Dependent variable: Log of bilateral export relative to the product of GDP in two countries

	(1)	(2)	(3)	(4)	(5)
Industry	**341**	**351**	**369**	**371**	**372**
Log of bilateral distance (km)	−1.911*** (0.034)	−1.416*** (0.028)	−1.514*** (0.026)	−1.891*** (0.032)	−1.737*** (0.043)
Common border dummy variable	0.490*** (0.065)	0.773*** (0.068)	1.054*** (0.073)	0.555*** (0.065)	1.056*** (0.095)
Common currency dummy variable	0.180*** (0.052)	0.075 (0.048)	−0.046 (0.050)	0.240*** (0.067)	0.262*** (0.076)
FTA dummy variable	0.217* (0.114)	−0.025 (0.113)	0.302*** (0.104)	−0.018 (0.158)	−0.330** (0.160)
ct1	0.122** (0.055)	0.033 (0.039)	−0.174*** (0.049)	0.148** (0.058)	0.041 (0.062)
ct2	0.026 (0.042)	0.017 (0.044)	−0.060 (0.047)	0.004 (0.049)	0.081 (0.060)
ct3	−0.449*** (0.068)	−0.057 (0.063)	0.041 (0.071)	0.025 (0.078)	0.049 (0.094)
ees1	0.055 (0.085)	0.109** (0.047)	−0.224*** (0.061)	0.071 (0.065)	−0.111 (0.090)
ees2	0.020 (0.080)	−0.034 (0.045)	−0.129** (0.063)	−0.075 (0.067)	−0.107 (0.094)
ees3	0.011 (0.085)	0.150*** (0.055)	−0.177** (0.063)	−0.022 (0.072)	0.042 (0.097)
Constant	−19.855*** (0.410)	−23.517*** (0.322)	−24.426*** (0.321)	−19.726*** (0.372)	−19.371*** (0.473)

(continued)

TABLE 3B

Results from the Competitiveness Analysis: Effects of Climate Change Measures on Energy-Intensive Industries *(continued)*

Industry	(1) 341	(2) 351	(3) 369	(4) 371	(5) 372
Exporting country fixed effects	Yes	Yes	Yes	Yes	Yes
Importing country fixed effects	Yes	Yes	Yes	Yes	Yes
Year fixed effects	Yes	Yes	Yes	Yes	Yes
Sample size	10,918	11,383	10,635	10,979	10,525
R-squares	0.7666	0.7265	0.7221	0.7085	0.6179

Note: *, **, *** indicate statistical significant at 10%, 5%, and 1% level, respectively.

Standard errors in parentheses are clustered by country-year pair.

TABLE 3C

Results from the Competitiveness Analysis: Effects of Climate Change Measures on Industries Subject to Higher Efficiency Standards

Dependent variable: Log of bilateral export relative to the product of GDP in two countries

Industry	(1) 381	(2) 382	(3) 383	(4) 384	(5) 385
Log of bilateral distance (km)	-1.389*** (0.022)	-1.112*** (0.021)	-1.171*** (0.024)	-1.313*** (0.029)	-0.937*** (0.020)
Common border dummy variable	0.883*** (0.049)	0.630*** (0.055)	0.502*** (0.058)	0.646*** (0.068)	0.947*** (0.064)
Common currency dummy variable	-0.041 (0.048)	-0.076 (0.053)	-0.066 (0.050)	-0.091* (0.053)	-0.032 (0.051)
FTA dummy variable	0.747*** (0.080)	0.628*** (0.081)	1.537*** (0.117)	1.482*** (0.126)	0.345*** (0.102)
ct1	0.003 (0.044)	-0.112*** (0.040)	0.066 (0.043)	-0.118** (0.054)	0.040 (0.040)
ct2	-0.013 (0.036)	0.014 (0.035)	-0.077* (0.040)	-0.016 (0.054)	0.159*** (0.044)
ct3	-0.273*** (0.060)	-0.369*** (0.061)	-0.464*** (0.066)	-0.439*** (0.082)	-0.258*** (0.061)
ees1	-0.307*** (0.054)	-0.050 (0.046)	0.027 (0.048)	-0.251*** (0.072)	-0.015 (0.058)
ees2	-0.082* (0.050)	-0.054 (0.042)	-0.018 (0.045)	-0.137** (0.067)	0.041 (0.056)
ees3	-0.214*** (0.057)	0.005 (0.047)	0.039 (0.053)	-0.242*** (0.068)	0.036 (0.060)
Constant	-24.224*** (0.266)	-25.087*** (0.255)	-25.925*** (0.291)	-24.286*** (0.373)	-27.934*** (0.234)
Exporting country fixed effects	Yes	Yes	Yes	Yes	Yes

TABLE 3C

Results from the Competitiveness Analysis: Effects of Climate Change Measures on Industries Subject to Higher Efficiency Standards *(continued)*

Industry	(1) 381	(2) 382	(3) 383	(4) 384	(5) 385
Importing country fixed effects	Yes	Yes	Yes	Yes	Yes
Year fixed effects	Yes	Yes	Yes	Yes	Yes
Sample size	11,568	11,742	11,602	11,272	11,451
R-squares	0.7667	0.7663	0.746	0.6307	0.7412

Note: *, **, *** indicate statistical significant at 10%, 5%, and 1% level, respectively.

Standard errors in parentheses are clustered by country-year pair.

APPENDIX 4

Industry-Specific Effects of Carbon Taxes and Energy Efficiency Standards

THE ENERGY-INTENSIVE INDUSTRIES CONSIDERED are paper and paper products (ISIC 341), industrial chemicals (351), nonmetallic products (369), iron and steel (371), and nonferrous metal (372). These are industries that should generally be adversely affected by a carbon tax. However, most governments also actively subsidize or exempt these industries to neutralize such adverse effects. Therefore, one may not be able to identify the impact of a carbon tax on these industries. Results, as summarized in table 4A, show that a carbon tax affects the paper and paper products industry (341) and the nonmetallic products industry (369).

For the nonmetallic mineral industry (such as cement), trade competitiveness is adversely affected when only the exporting country imposes the tax, although it is not affected when both countries impose the tax. This suggests that in the case of a nonmetallic industry such as the cement industry, a unilateral domestic environmental measure hurts the export performance of the country. This argument is used by a number of governments in order to justify direct subsidies to these industries to offset the adverse shock of a carbon tax.

On the other hand, for the paper and paper products industry, trade competitiveness actually improves if the exporting countries impose the tax. This indicates that the governments may have overly subsidized this industry, which causes the expansion in trade. Interestingly, when both importing and exporting countries have a carbon tax, it leads to a reduction in paper trade. Another industry that

TABLE 4A
Impact of Carbon Taxes and Energy Efficiency Standards on Export Competitiveness (Energy-Intensive Industries)

	Carbon Tax (Imposed by Country)			Energy Efficiency Standards (Imposed by Country)		
Industry	Exporting	Importing	Exporting and Importing	Exporting	Importing	Exporting and Importing
Paper and Paper Products (341)	Significant (+)		Highly Significant (–)			
Industrial Chemicals (351)				Significant (+)		Highly Significant (+)
Nonmetallic Mineral Products (369)	Highly Significant (–)			Highly Significant (–)	Significant (–)	Highly Significant (–)
Iron and Steel (371)	Significant (+)					
Nonferrous Metal (372)						

Note: (–) denotes a decrease in trade and (+) denotes an increase in trade.

TABLE 4B
Impact of Carbon Taxes and Energy Efficiency Standards on Export Competitiveness (Industries Subject to Higher Energy Efficiency Standards)

	Carbon Tax (Imposed by Country)			Energy Efficiency Standards (Imposed by Country)		
Industry	Exporting	Importing	Exporting and Importing	Exporting	Importing	Exporting and Importing
Metal Products (381)			Highly Significant (–)	Highly Significant (–)	Marginally Significant (–)	Highly Significant (–)
Machinery (382)	Highly Significant (–)		Highly Significant (–)			
Electrical Machinery (383)		Marginally Significant (–)	Highly Significant (–)			
Transport Equipment (384)	Significant (–)		Highly Significant (–)	Highly Significant (–)	Significant (–)	Highly Significant (–)
Scientific Equipment (385)		Highly Significant (–)	Highly Significant (–)			

Note: (–) denotes a decrease in trade and (+) denotes an increase in trade.

also may have benefited from carbon tax due to government subsidies is the iron and steel industry (371), where trade increases when only exporting countries impose the tax.

A very different picture emerges when the focus is on those industries that produce outputs that are subjected to higher energy efficiency standards. Industries that are usually subjected to higher energy efficiency standards are metal products (ISIC 381), machinery (382), electrical machinery (383), transport equipment (384), and scientific equipment (385). Here, one expects energy efficiency standards to have a negative impact on trade. Results confirm that only some of these industries are adversely affected by the standards requirements, and the effects are particularly large for metal products (381) and transport equipment (384). In both these industries, it does not matter whether such a standard requirement is imposed by the exporting country or the importing country or both: the trade is reduced by 20 to 30 percent.

Perhaps the most interesting finding of table 4B is that these industries are also adversely affected by a carbon tax. Bilateral trade in some cases, such as in the electronic industry, is reduced by as much as 40 percent. This seems to suggest that industries that are not usually exempted or subsidized by the governments may be bearing the brunt of the (negative) impact of the carbon tax. This also possibly indicates that some third countries that do not have a carbon tax may be benefiting overall from the situation when both exporting and importing countries impose the tax.

APPENDIX 5

Partial Equilibrium Trade Policy Simulation Model

ONE OF THE KEY EFFECTS SIMULATED in standard partial equilibrium trade models (see Laird and Yeats 1990 for details) is trade creation; that is, the increased demand in country j for commodity i resulting from the price response when tariffs are reduced or eliminated. In the case where product i faces a tariff, the partial equilibrium approach starts with the assumption that the percentage change in imports of (dM/Mij) can be derived from:

$$dMij/Mij = ed * (dPij/Pij) \tag{1}$$

where ed is the elasticity of import demand for i and (dPij/Pij) is the percentage change in the price of the product resulting from the tariff cut. Manipulation of the terms in equation (1), and also assuming a non-zero elasticity of supply (es) allows one to directly estimate trade creation (TCij) from the following:

$$TCij = Mij * ed * dt/((l + ti)\ (1 - ed/es)) \tag{2}$$

where Mij represents the initial level of imports before the tariff cut and t is the initial import tariff. The crucial link between equation (1) and (2) is that the percentage change in price due to the tariff (dPij/Pij) is assumed equal to the term $(dt/((l + ti)\ (1 - ed/es)))$.

If an infinite elasticity of supply is assumed, the equation (2) is reduced to:

$$TCij = Mij * ed * dt/(l + ti) \tag{3}$$

To account for the tendency of importers to substitute goods from one source to another due to a change in relative prices, one can estimate trade diversion from the following equation:

$$TDii = TCij\ (Mij/Vij) \qquad (4)$$

where (Mij/Vij) is the import penetration ratio or the share of imports from non-preference-receiving countries in domestic consumption of the product.

APPENDIX 6

Maximum and Applied Tariff Rates on Select Climate-Friendly Technologies

HS Code	Product Description	Low- and Middle-Income WTO Members		High-Income WTO Members	
		Maximum Average Bound Tariffs	Average Applied Tariff Rates	Maximum Average Bound Tariffs	Average Applied Tariff Rates
392010	PVC or polyethylene plastic membrane systems to provide an impermeable base for landfill sites and protect soil under gas stations, oil refineries, etc. from infiltration by pollutants and for reinforcement of soil	30	13	15	5
560314	Nonwovens, whether or not impregnated, coated, covered or laminated: of manmade filaments; weighing more than 150 g/m^2 for filtering wastewater	33	14	16	4
701931	Thin sheets (voiles), webs, mats, mattresses, boards, and similar nonwoven products	34	13	17	4
730820	Towers and lattice masts for wind turbine	28	10	16	3
730900	Containers of any material, of any form, for liquid or solid waste, including for municipal or dangerous waste	32	12	17	4
732111	Solar driven stoves, ranges, grates, cookers (including those with subsidiary boilers for central heating), barbecues, braziers, gas-rings, plate warmers and similar non-electric domestic appliances, and parts thereof, of iron or steel	36	18	15	5
732190	Stoves, ranges, grates, cookers (including those with subsidiary boilers for central heating), barbecues, braziers, gas-rings, plate warmers and similar non-electric domestic appliances, and parts thereof, of iron or steel—Parts	36	14	15	4
732490	Water saving shower	28	19	17	4
761100	Aluminum reservoirs, tanks, vats and similar containers for any material (specifically tanks or vats for anaerobic digesters for biomass gasification)	31	11	16	4
761290	Containers of any material, of any form, for liquid or solid waste, including for municipal or dangerous waste	31	13	14	4
840219	Vapor generating boilers, not elsewhere specified or included hybrid	24	5	15	4
840290	Super-heated water boilers and parts of steam generating boilers	21	5	15	4

840410	Auxiliary plant for steam, water, and central boiler	25	5	15	3
840490	Parts for auxiliary plant for boilers, condensers for steam, vapor power unit	25	4	16	3
840510	Producer gas or water gas generators, with or without purifiers	24	5	13	2
840681	Turbines, steam and other vapor, over 40 MW, not elsewhere specified or included	28	5	13	3
841011	Hydraulic turbines and water wheels of a power not exceeding 1,000 kW	24	4	15	3
841090	Hydraulic turbines and water wheels; parts, including regulators	24	4	15	3
841181	Gas turbines of a power not exceeding 5,000 kW	20	5	13	2
841182	Gas turbines of a power exceeding 5,000 kW	20	5	13	2
841581	Compression type refrigerating, freezing equipment incorporating a valve for reversal of cooling/heating cycles (reverse heat pumps)	29	13	16	4
841861	Compression type refrigerating, freezing equipment incorporating a valve for reversal of cooling/heating cycles (reverse heat pumps)	21	7	17	4
841869	Compression type refrigerating, freezing equipment incorporating a valve for reversal of cooling/heating cycles (reverse heat pumps)	21	7	16	4
841919	Solar boiler (water heater)	27	10	17	4
841940	Distilling or rectifying plant	23	4	15	3
841950	Solar collector and solar system controller, heat exchanger	24	5	15	3
841989	Machinery, plant or laboratory equipment whether or not electrically heated (excluding furnaces, ovens etc.) for treatment of materials by a process involving a change of temperature such a heating, cooking, roasting, distilling, rectifying, sterilizing, steaming, drying, evaporating, vaporizing, condensing or cooling.	25	6	12	3
841990	Medical, surgical or laboratory stabilizers	24	6	12	2
848340	Gears and gearing and other speed changers (specifically for wind turbines)	22	8	16	3
848360	Clutches and universal joints (specifically for wind turbines)	23	9	15	3

(continued)

HS Code	Product Description	Low- and Middle-Income WTO Members		High-Income WTO Members	
		Maximum Average Bound Tariffs	Average Applied Tariff Rates	Maximum Average Bound Tariffs	Average Applied Tariff Rates
850161	AC generators not exceeding 75 kVA (specifically for all electricity generating renewable energy plants)	27	7	15	3
850162	AC generators exceeding 75 kVA but not 375 kVA (specifically for all electricity generating renewable energy plants)	26	7	16	3
850163	AC generators not exceeding 375 kVA but not 750 kVA (specifically for all electricity generating renewable energy plants)	26	5	16	3
850164	AC generators exceeding 750 kVA (specifically for all electricity generating renewable energy plants)	28	5	16	3
850231	Electric generating sets and rotary converters; wind-powered	26	5	16	3
850680	Fuel cells use hydrogen or hydrogen-containing fuels such as methane to produce an electric current, through an electrochemical process rather than combustion	25	18	16	3
850720	Other lead acid accumulators	24	16	16	5
853710	Photovoltaic system controller	26	10	17	3
854140	Photosensitive semiconductor devices, including photovoltaic cells whether or not assembled in modules or made up into panels; light-emitting diodes	21	4	9	1
900190	Mirrors of other than glass (specifically for solar concentrator systems)	30	7	16	3
900290	Mirrors of glass (specifically for solar concentrator systems)	29	12	18	3
903210	Thermostats	33	7	14	3
903220	Manostats	33	6	13	2

Bibliography

Alavi, R. 2007. "An Overview of Key Markets, Tariffs and Non-tariff Measures on Asian Exports of Select Environmental Goods." ICTSD Trade and Environment Series Issue Paper 4, International Centre for Trade and Sustainable Development, Geneva.

Barbier, E. B., R. Damania, and D. Léonard. 2005. "Corruption, Trade and Resource Conversion." *Journal of Environmental Economics and Management* 50: 276–99.

Baumert, K., and N. Kete. 2002. "Introduction: An Architecture for Climate Protection." In *Building on the Kyoto Protocol: Options for Protecting the Climate,* ed. K. Baumert et al. Washington, DC: World Resources Institute.

Baumol, W. J., and W. E. Oates. 1988. *The Theory of Environmental Policy.* 2nd ed. New York: Cambridge University Press.

Bhagwati, J., and P. Mavroidis. 2007. "Is Action Against U.S. Exports for Failure to Sign Kyoto Protocol WTO-Legal?" *World Trade Review* (Cambridge Journals) 6: 299–310.

Biermann, F., and R. Brohm. 2003. "Implementing the Kyoto Protocol Without the United States: The Strategic Role of Energy Tax Adjustments at the Border." Global Governance Working Paper 5, Global Governance Project, Potsdam, Berlin, Oldenburg. http://www.glogov.org.

Brewer, T. L. 2003. "The Trade Regime and the Climate Regime: Institutional Evolution and Adaptation." *Climate Policy* 3 (4): 329–41.

———. 2004. "The WTO and the Kyoto Protocol: Interaction Issues." *Climate Policy* 4 (1): 3–12.

———. 2007. "Climate Change Technology Transfer: International Trade and Investment Policy Issues in the G8+5 Countries." Paper prepared for G8+5 Climate Change Dialogue. (Available from the author, e-mail brewert@georgetown.edu).

Buccini, J. 2004. *The Global Pursuit of the Sound Management of Chemicals.* Washington, DC: World Bank.

CESTT (Centre for Environmentally Sound Technology Transfer). 2002. *Doing Business in the Chinese Environmental Market.* Beijing: CESTT.

Charnovitz, S. 2003. *Trade and Climate: Potential for Conflicts and Synergies.* Washington, DC: Pew Center on Global Climate Change.

Commission of the European Communities. 2005. "Report on Demonstrable Progress Under the Kyoto Protocol." COM (2005) 615, CEC, Brussels.

Copeland, B. R. 1996. "Pollution Content Tariffs, Environmental Rent Shifting, and the Control of Cross-Border Pollution." *Journal of International Economics* 40 (3): 459–76.

Copeland, B. R., and M. S. Taylor. 2004. "Trade, Growth, and the Environment." *Journal of Economic Literature* 42: 7–71.

Cosbey, A. 2003. "The Kyoto Protocol and the WTO." Seminar note for the Energy and Environment Program, Royal Institute of International Affairs, London.

Eskom. 2005. "National Efficient Lighting: Roll-out Initiative." Eskom, Johannesburg, SA. http://www.eskomdsm.co.za/pdfs/NELflyer_engl_afrikaans.pdf.

European Energy Agency. 2004. "Analysis of Greenhouse Gas Emission Trends and Projections in Europe 2004," EEA Technical Report 7/2004. http:// reports.eea.europa.eu/technical_report_2004_7/en.

Feenstra, R. 2003. *Advanced International Trade: Theory and Evidence.* Princeton, NJ: Princeton University Press.

Fliess, B. A., and P. Sauve. 1997. "Of Chips, Floppy Disks and Great Timing: Assessing the Information Technology Agreement." Paper prepared for the Institute Francais des Relations Internationales (IFRI) and the Tokyo Club Foundation for Global Studies.

Fouquet, D., and T. Johansson. 2005. "Energy and Environmental Tax Models from Europe and Their Link to Other Instruments for Sustainability: Policy Evaluation and Dynamics of Regional Integration." Presentation at the Eighth Senior Policy Advisory Committee Meeting, Beijing, China, November 18.

Frankel, J. 2003. "The Environment and Globalization." NBER Working Paper 10090, National Bureau of Economic Research, Cambridge, MA.

———. 2004. "Kyoto and Geneva: Linkage of the Climate Change and the Trade Regime." Paper prepared for the "Broadening Climate Change Discussion: The Linkage of Climate Change to Other Policy Areas," FEEM/MIT, Venice, Italy.

Fredriksson, P. G., and M. Mani. 2004. "Trade Integration and Political Turbulence: Environmental Policy Consequences." *Advances in Economic Analysis and Policy, The Pollution Haven Hypothesis* 4 (2).

Fredriksson, P. G., H. R. J. Vollebergh, and E. Dijkgraaf. 2004. "Corruption and Energy Efficiency in OECD Countries: Theory and Evidence." *Journal of Environmental Economics and Management* 47 (2): 207–31.

GEF (Global Environment Facility). 2004. IFC/GEF Efficient Lighting Initiative (office memorandum), GEF, Washington, DC.

Georgieva, K. and M. Mani. 2006. "Trade and the Environment Debate: WTO, Kyoto and Beyond." In David Tarr, ed.,*Trade Policy and WTO Accession for Economic Development in Russia and the CIS: A Handbook* [Russian]. Moscow: Ves Mir.

Hamwey, R. 2005. "Environmental Goods: Where Do the Dynamic Trade Opportunities for Developing Countries Lie?" Cen2eco Working Paper, Centre for Economic and Ecological Studies, Geneva. http://www.cen2eco.org/C2E-Documents/Cen2eco-EG-DynGains-W.pdf.

Harris, M. N., L. Kónya, and L. Mátyás. 2002. "Modeling the Impact of Environmental Regulations on Bilateral Trade Flows: OECD, 1990–1996." *The World Economy* 25 (3): 387–405.

ICTSD (International Centre for Trade and International Development). 2006. "Developing Countries Present Views on Environmental Goods." *Bridges Weekly* 10 (22). http://www.ictsd.org/weekly/06-06-21/story6.htm.

———. 2007a. *Illustrative Analysis of Sectors Relevant to Air-pollution and Renewable Energy.* ICTSD Trade and Environment Series Issue Paper 6, International Centre for Trade and Sustainable Development, Geneva.

———. 2007b. "NAMA: State of Suspended Pessimism." *Bridges Monthly Review* 11 (1). http://www.ictsd.org/monthly/bridges/BRIDGES11-1.pdf.

IEA (International Energy Agency). 1999. Energy Policies of IEA Countries: A Review. Paris: OECD/IEA.

———. 2001. *Dealing with Climate Change: Policies and Measures in IEA Member Countries.* Paris: OECD/IEA.

———. 2002. *Energy Efficiency Update.* Paris: OECD/IEA.

———. 2006. *World Energy Outlook 2006.* Paris: IEA.

IETA (International Emissions Trading Association). 2005. "Japan Launches Voluntary Emissions Trading Scheme." News Archive 2005 (September 28). IETA, Geneva. http://www.ieta.org/ieta/www/pages/index.php?IdSitePage=962.

India's Ministry of Non-conventional Energy Sources. 2004. Renewable Energy in India – Business Opportunities. "Fiscal Incentives." http://mnes.nic.in. http://mnes.nic.in/business%20oppertunity/index.htm.

IPCC (Intergovernmental Panel on Climate Change). 2001. *Climate Change 2001: Mitigation.* Working Group III contribution to the IPCC, 3rd Assessment Report. http:// www.grida.no/climate/ipcc_tar/wg3/index.htm.

———. 2007. *Climate Change 2007: Mitigation of Climate Change: Summary for Policymakers.* Working Group III contribution to the IPCC, 4th Assessment Report. http:// www.ipcc.ch/SPM040507.pdf.

Jin, Y., and X. Liu. 1999. "Clean Coal Technology Acquisition: Present Situation, Obstacles, Opportunities, and Strategies for China." Working Group on Trade and Environment. The Third Meeting of the Second Phase of CCICED: China Council for International Cooperation on Environment and Development, Guanghua School of Management of Peking University, Beijing.

Kee, H. L., A. Nicita, and M. Olarreaga. 2005a. "Estimating Trade Restrictiveness Indices." Policy Research Working Paper Series 3840, World Bank, Washington, DC.

———. 2005b. "Import Demand Elasticities and Trade Distortions," Policy Research Working Paper Series 3452, World Bank, Washington, DC.

Kim, J. 2007. "Issues of Dual-Use and Reviewing Product Coverage of Environmental Goods." OECD Trade and Environment Working Paper 2007-01, OECD, Paris.

Kojima, M., D. Mitchell, and William Ward. 2006. *Considering Trade Policies for Liquid Biofuels.* Washington, DC: World Bank.

Laird, S., and A. Yeats. 1990. *Quantitative Methods for Trade-Barrier Analysis.* New York: New York University Press.

Lecocq, F., and Z. Shalizi. 2004. "Will Kyoto Protocol Affect Growth in Russia?" Policy Research Working Paper 3454, World Bank, Washington, DC.

Lefèvre, N., P. T'Serclaes, and P. Waide, 2006. "Barriers to Technology Diffusion: The Case of Compact Flourescent Lamps." Joint OECD and IEA Paper, Paris.

Linnemann, H. 1966. *An Econometric Study of International Trade Flows.* Amsterdam: North-Holland.

Ludema, R., and I. Wooten. 1994. "Cross-border Externalities and Trade Liberalization: The Strategic Control of Pollution." *Canadian Journal of Economics* 27: 950–66.

Mani, M. 1996. "Environmental Tariffs on Polluting Imports: An Empirical Study." *Environmental and Resource Economics* 7 (4): 391–412.

Mani, M., and D. Wheeler. 1998. "In Search of Pollution Havens? Dirty Industry in the World Economy, 1960–1995." *Journal of Environment and Development* 7 (3): 215–47.

Markusen, J. R. 1975. "International Externalities and Optimal Tax Structures." *The Journal of International Economics* 5: 15–29.

Mytelka, L. 2007. "Technology Transfer Issues in Environmental Goods and Services", draft. International Centre for Trade and Sustainable Development, Geneva.

Nicita, A., and M. Olarreaga. 2004. "Exports and Information Spillovers." Policy Research Working Paper 2474, World Bank, Washington, DC.

Nordhaus, W. D. 2007. "To Tax or Not to Tax: Alternative Approaches to Slowing Global Warming." *Review of Environmental Economics and Policy 2007* 1 (1): 26–44.

OECD (Organisation for Economic Co-operation and Development). 1997. *Economic/Fiscal Instruments: Taxation (i.e., Carbon/Energy).* Paris: OECD.

———. 1999. *Sustainable Economic Growth: Natural Resources and the Environment in Norway.* Paris: OECD.

———. 2001. *Environmentally Related Taxes in the OECD Countries.* Paris: OECD.

———. 2002. *Foreign Direct Investment and Development. Maximizing Benefits, Minimizing Costs.* Paris: OECD.

———. 2003. *Policies to Reduce Greenhouse Gas Emissions in Industry – Successful Approaches and Lessons Learned.* Paris: OECD.

———. 2005. *Policies to Reduce Greenhouse Gas Emissions in Industry: Implications for Steel.* Paris: OECD.

———. 2006. *The Political Economy of Environmentally Related Taxes.* Paris: OECD.

Pacala, S. W., and R. H. Socolow. 2004. "Stabilization Wedges: Solving the Climate Problem for the Next 50 Years with Current Technologies." *Science* 305: 968–72.

Pauwelyn, J. 2007. "U.S. Federal Climate Policy and Competitiveness Concerns: The Limits of International Trade Law." NI WP 07-02. April.

Petsonk, A. 1999. "The Kyoto Protocol and the WTO: Integrating Greenhouse Gas Emissions Allowance Trading into the Global Marketplace." *Duke Environmental Law and Policy Forum* 10.

REN21 (Renewable Energy Policy Network for the 21st Century). 2005. *Renewables 2005 Global Status Report.* Washington, DC: Worldwatch Institute.

REN21. 2006. *Renewable Energy Global Status Report 2006 Update.* Paris: REN 21 Secretariat; Washington, DC: Worldwatch Institute.

Sands, P. 2005. *Lawless World: America and the Making and Breaking of Global Rules.* London: Penguin Books.

Socolow, R., R. Hotinski, J. B. Greenblatt, and S. Pacala. 2004. "Solving the Climate Problem: Technologies Available to Curb CO_2 Emissions." *Environment* 46 (10): 8–19.

Steenblik, R. 2005. "Liberalisation of Trade in Renewable-Energy Products and Associated Goods; Charcoal, Solar Photovoltaic System, and Wind Pumps and Turbines." Trade and Environment Working Paper 2005-07, OECD, Paris.

Stern, N. 2006. *Stern Review on the Economics of Climate Change.* London: HM Treasury, U.K.

Stiglitz, J. E. 2006. *Making Globalization Work.* New York: WW Norton.

Sugathan, M., E. Claro, N. Lucas, and M. Marconini. Forthcoming. "Environmental Goods and Services: A Compendium on Domestic and WTO Negotiating Strategy." Trade and Environment Series Issue Paper. TD/B/COM.1/EM.21/CRP.1, ICTSD, Geneva.

Swiss Confederation to the COP. 2005. *Switzerland's Report on Demonstrable Progress in Line with Decisions 22/CP.7 and 25/CP.8 of the UNFCCC.* Zurich: SAEFL.

Tinbergen, J. 1962. *Shaping the World Economy: Suggestions for an International Economic Policy.* New York: The Twentieth Century Fund.

UNCTAD (United Nations Conference on Trade and Development). 1995. *Environmentally Preferable Products (EPPs) as a Trade Opportunity for Developing Countries.* UNCTAD/COM/70. Geneva: UNCTAD.

———. 2003. "Environmental Goods: Trade Statistics of Developing Countries." United Nations, Comtrade Database.

UNFCCC (United Nations Framework Convention on Climate Change). 2005. "Report on the Round-Table Discussion on Experiences of Parties Included in Annex I to the Convention in Implementing Policies and Measures." Presented at the Subsidiary Body for Scientific and Technological Advice, 23rd session, November 28–December 6, 2005, Montreal.

USTR (U.S. Trade Representative). 2006. *Report by the Office of the United States Trade Representative on Trade-Related Barriers to the Export of Greenhouse Gas Intensity Reducing Technologies.* Washington, DC: USTR.

van Beers, C., and A. de Moor 1998. "Scanning Subsidies and Policy Trends in Europe and Central Asia." NEP/DEIA/TR.98-2. Nairobi: UNEP.

van Beers, C., and J. van den Bergh. 1997. "Empirical Multi-Country Analysis of the Impact of Environmental Regulations on Foreign Trade Flows." *Kyklos* 50: 29–46.

———. 2001. "Perseverance of Perverse Subsidies and Their Impact on Trade and Environment." *Ecological Economics* 36: 475–86.

Vernstrom, R. 2007. "Renewable Energy Tax Rationalization: An Assessment of Import Tax Policy in Cambodia." Consultant report prepared for the World Bank, Washington, DC.

Vikhlyaev, A. 2003. "Environmental Goods and Services: Defining Negotiations or Negotiating Definitions?" *Trade and Environment Review 2003.* Geneva: UNCTAD.

Waide, P., and M. K. Gueye. 2007. "Scaling Up Energy Efficiency: The Problem of Market Access." *Bridges Monthly Review* 11 (3). http://www.ictsd.org.

Werksman, J. 1999. "Greenhouse Gas Emissions Trading and the WTO." *Review of European Community and International Environmental Law* 8 (3).

Wilder, M., M. Willis, and P. Curnow. 2006. *The Clean Development Mechanism: Special Considerations for Renewable Energy Projects.* Chicago: Baker & McKenzie.

World Bank. 1992. "International Trade and the Environment," ed. Patrick Low, World Bank Discussion Paper 159, World Bank, Washington, DC.

———. 1999. "Trade, Global Policy and Environment." World Bank Discussion Paper 402, World Bank, Washington, DC.

———. 2006a. *An Investment Framework for Clean Energy and Development: A Progress Report, Submitted to the Development Committee.* Joint Ministerial Committee of the Boards of Governors of the Bank and the Fund on the Transfer of Real Resources to Developing Countries, September 5.

———. 2006b. *World Development Indicators.* Washington, DC: World Bank.

———. 2007a. *Clean Energy and Development: Toward an Investment Framework.* Washington, DC: World Bank.

———. 2007b. *Global Economic Prospects 2007: Managing the Next Wave of Globalization.* Washington, DC: World Bank.

World Bank and IETA (International Emissions Trading Association). 2006. *State and Trends of the Carbon Market 2006.* Washington, DC: World Bank; Geneva: IETA.

WTO (World Trade Organization). 1999. *Trade and Environment.* Geneva: WTO.

———. 2002. *List of Environmental Goods.* Doc.TN/TE/W/18. Geneva: WTO.

———. 2004. "Trade and Environment at the WTO." Background paper, WTO, Geneva.

———. 2007. "Lamy Says ITA Success Is Inspiration to Doha Negotiators." WTO 2007 News. http://www.wto.org/english/news_e/news07_e/symp_ita_march07_e.htm, March 28, 2007.

Zhang, Z., and L. Assuncao. 2004. "Domestic Climate Policies and the WTO." *The World Economy* 27 (3).

INDEX

F

G

H

ECO-AUDIT

Environmental Benefits Statement

The World Bank is committed to preserving endangered forests and natural resources. The Office of the Publisher has chosen to print ***International Trade and Climate Change: Economic, Legal, and Institutional Perspectives*** on recycled paper with 100% post-consumer waste. The Office of the Publisher has agreed to follow the recommended standards for paper usage set by the Green Press Initiative, a nonprofit program supporting publishers in using fiber that is not sourced from endangered forests. For more information, visit www.greenpressinitiative.org.

Saved:

- 26 trees
- 18 million BTUs of total energy
- 2,295 lbs. of net greenhouse gases
- 9,525 gallons of waste water
- 1,223 lbs. of solid waste